"Oreskes joins a distinguished line of thinkers
trust the findings of the scientific community. . . . Oreskes clearly reminds
readers that science has consistently brought home the bacon."
—*Kirkus Reviews*

"A fascinating and accessible read that considers numerous domains and
issues to bring the reader to Oreskes' ultimate point, that trustworthy
science depends on consensus, diversity, and methodological
openness and flexibility."
—**JEFF SHARE,** *Journal of Sustainability Education*

"This book is well worth the effort for anyone concerned about climate
change, protection of biodiversity, and other issues that involve science
advising policy. Insights from Naomi Oreskes can bolster our arguments
countering the anti-science, anti-expertise, anti-intellectual forces
at work in the world today."
—**JOHN MILES,** *National Parks Traveler*

"A marvellous, up to date, thorough historical survey of science and
its processes."
—**JOHN R. HELLIWELL,** *Journal of Applied Crystallography*

"How do we get to the truth? How do we safeguard scientific knowledge
(and ourselves) from those whose interests are threatened by it? With her
trailblazing work on climate denial and much else, Naomi Oreskes offers
essential perspective on these questions. She tackles them head-on
in this clear, utterly compelling book."
—**NAOMI KLEIN,** author of *No Is Not Enough* and *This Changes Everything*

"This comprehensive and thoughtful book explores the thorny questions
we often take for granted regarding why, when, and how we can—or
can't—trust science. In a post-truth world, this is the book we need."
—**KATHARINE HAYHOE,** Texas Tech University, coauthor of
A Climate for Change

"In an age of fake news, alternative facts, and the notion that opinion and
ideology trump empirical evidence and the scientific method, how should
science respond? The title of this incredibly important book poses one of
the most urgent questions of our time, because if we don't trust science
then humanity is doomed."
—**JIM AL-KHALILI, FRS**, physicist, author, and host of BBC's
The Life Scientific

"Anybody who wants to understand the conceptual and practical under-
pinnings of credibility in scientific findings should read this book."
—**JOHN P. HOLDREN,** Harvard University, former science and
technology adviser to President Barack Obama

"This is a troubled time in the history of science and a perilous one for its reputation with the public, which is why now is exactly the right time for the fearless and brilliant Naomi Oreskes to explore this issue. The result is a don't-miss investigation into the very human nature of research—its successes, its failures, and its fundamental integrity in the search for truth."
—**DEBORAH BLUM,** Pulitzer Prize–winning author of
The Poison Squad

"*Why Trust Science?* is a timely book by one of the world's most important and trenchant observers of science and society. With misinformation and disinformation rampant today, caring citizens do not know what or whom to trust and have become confused about evidence, opinion, and partisan assertion. We need Oreskes's clear look at how to recognize and use reliable knowledge. I cannot overstate the importance of this book now to scientists and citizens."
—**RUSH D. HOLT,** CEO of the American Association for the Advancement of Science, former US House member

"Naomi Oreskes, who grabbed our attention with her keen insights into climate denial, now tackles a threat to the very basis of an informed democracy—attacks on science itself. Captivating, forceful, and grounded in critical analysis, *Why Trust Science?* is for anyone who cares about our world."
—**JANE LUBCHENCO,** former head of the National Oceanic and Atmospheric Administration

"In this authoritative defense of science, noted scientist and science historian Naomi Oreskes presents her case, subjects it to scrutiny by experts, and responds to the points raised. Her approach itself is a metaphor for the self-correcting machinery of science and the iterative process that leads science toward a better understanding of the natural world."
—**MICHAEL E. MANN,** Penn State University, author of *The Hockey Stick and the Climate Wars*

"This book poses an important and timely question. While acknowledging the ways that science can go off track and become unreliable, Oreskes provides a compelling and well-supported defense of science, arguing that its trustworthiness derives from its collective character rather than a particular method or the inherent objectivity of scientists."
—**ANGELA N. H. CREAGER,** author of *Life Atomic: A History of Radioisotopes in Science and Medicine*

"An insightful, lucid, and accessible discussion of a highly complex issue of great urgency and importance. Oreskes's call for a socially engaged science might lead to substantial changes in our conception of the role of science in society and the ways in which science is organized institutionally."
—**KARIM BSCHIR,** ETH Zürich

WHY
TRUST
SCIENCE
?

The University Center for Human Values Series

Stephen Macedo, Editor

WHY TRUST SCIENCE

?

NAOMI ORESKES

With a new preface by the author

PRINCETON UNIVERSITY PRESS

PRINCETON AND OXFORD

Published by Princeton University Press
41 William Street, Princeton, New Jersey 08540
6 Oxford Street, Woodstock, Oxfordshire OX20 1TR

press.princeton.edu

LCCN 2020949715
First paperback edition, with a new preface by the author, 2021
Paperback ISBN 9780691212265
Cloth ISBN 9780691179001

British Library Cataloging-in-Publication Data is available

Editorial: Alison Kalett and Kristin Zodrow
Production Editorial: Sara Lerner
Text Design: Leslie Flis
Cover Design: Amanda Weiss
Production: Jacqueline Poirier
Publicity: Sara Henning-Stout and Katie Lewis
Copyeditor: Brittany Micka-Foos

This book has been composed in Arno Pro with Gotham Display

Printed in the United States of America

Trust, but verify.

—RONALD REAGAN

CONTENTS

COVID-19. Rarely does the world offer proof of an academic argument, and even more rarely in a single word or term. But there it is. COVID-19 has shown us in the starkest terms—life and death—what happens when we don't trust science and defy the advice of experts.

As of this writing, the United States leads the world in both total cases and total deaths from COVID-19, the disease caused by the novel coronavirus that appeared in 2019. One might think that death rates would be highest in China, where the virus first emerged and doctors were presumably caught unprepared, but that is not the case. According to *The Lancet*—the world's premier medical journal—as of early October 2020, China had confirmed 90,604 cases of COVID-19 and 4,739 deaths, while the United States had registered 7,382,194 cases and 209,382 deaths.[1] And China has a population more than four times that of the United States. If the United States had a pandemic pattern similar to China, we would have seen only 22,500 cases and 1128 deaths.

While COVID-19 has killed people across the globe, death rates have been far higher in the United States than in other wealthy countries, such as Germany, Iceland, South Korea, New Zealand, and Taiwan, and even than in some much poorer countries, such as Vietnam.[2] The Johns Hopkins University School of Medicine puts the US death rate per 100,000 people at 65.5.[3] In Germany it is 11.6. In Iceland, 2.83. In South Korea, 0.89. In New Zealand, 0.51. In China, 0.34. And in Taiwan and Vietnam? 0.03 and 0.04. If the American death rate had been similar to New Zealand's, instead of seeing more than 200,000 deaths in the first ten months of the pandemic, we would have seen fewer than 2,000. If we were like Vietnam, we would have seen a little over 100.[4]

Death rate is an imperfect guide to a pandemic, because it is affected by many factors, including population structure, access to health care, and the underlying health of the population. Death rates are also affected by reporting and testing. A country like China, with low transparency, may not be reporting everything accurately. A metropolis like New York City, caught by surprise with inadequate testing capacity in the early stages of the pandemic, probably underestimated the number of cases and therefore overestimated the death rate. (This could help to explain why the death rate in New York appeared to be much higher than elsewhere in the United States.) And since COVID-19 is very deadly to the elderly, a country with an aged population can be expected to see a higher death rate than one with a younger population, but by that measure, Germany should have done more poorly than the United States. In fact, it has done far better.[5] Perhaps the most compelling statistic is this: the United States has 4% of the global population, and it has had 20% of global deaths.

By any measure, the US response has been a disaster. But rather than ask why it has been so bad, it may be more instructive to ask: What is common to the countries that have done well? The answer is straightforward: The countries that have seen low death rates effectively controlled the spread of the virus, and they did so by trusting science.

In December 2019, when COVID-19 first emerged, public health experts raised the alarm that we were seeing a novel virus—of "unknown etiology"—that could pose a pandemic threat.[6] By the end of January 2020, the World Health Organization declared the coronavirus outbreak a PHEIC—a public health emergency of international concern.[7] This was only the sixth time the WHO had invoked this measure since the regulations under which it operates were established in 2005.

Public health experts immediately made recommendations about how to minimize the disease spread. These included frequent, thorough hand washing with soap and hot water; avoiding large public gatherings; and staying home at the first sign of illness. Admittedly, these recommendations were not 100% consistent—this was, after all, a *novel* disease, so there was much about it that was unknown—and the WHO offered contradictory advice on masks. But this was not because the organization did not have reason to think that masks might help. It was because it was afraid that people would hoard them, exacerbating an already serious shortage of masks for health care and other essential workers.[8] (The WHO's confusing mask guidance—which it later altered—was not a failure of scientific knowledge but a failure of scientific communication, grounded in expert distrust of lay people. But this distrust—a better word might be "caution"—was perhaps warranted, given how many people did, in fact, hoard toilet paper, disinfectants, and other essential supplies.) Other scientists felt that in the absence of convincing scientific evidence that masks would work to stop *this particular virus*, they could not recommend the use of them.[9] Overall, however, most of the public health advisories were consistent, based on existing scientific knowledge of how respiratory viruses spread.[10]

In the United States, a great deal of attention has focused on individual action—hand washing, staying home, wearing masks—but public health officials also recommended measures that prior epidemics had proved effective: testing, isolation of sick individuals, contact tracing, and where needed, quarantine. These measures had helped in past pandemics and therefore had at least some likelihood of working in this one. (The word "quarantine," after all, is a very old one, dating from fourteenth-century Italy, where incoming ships were required to stay in port for forty days: *quaranta giorni*.)

More important, a broad program of testing, isolation, and contact tracing was scientific common sense, because viruses do not spread by magic; they spread from sick people to well ones. If you can quickly identify the sick and separate them from the healthy, then you have a good chance of reducing the spread. The countries that can today boast of very low caseloads and death rates all took this scientific experience and expertise to heart.

Vietnam is a case in point.[11] Early in the pandemic the government implemented strict measures to test any symptomatic person, and, where results were positive, to trace, test, and isolate their contacts. The government also promoted the use of mobile apps by which people could record their symptoms and get tested promptly as needed. Passengers arriving from overseas were quarantined, and in a few cases—such as a man returning from a religious festival in Malaysia—the government ordered targeted lockdowns, in this case of a mosque he had visited in Ho Chi Minh City and of his entire home province.[12] The government also restricted travel and public gatherings, and ordered the shutdown of many non-essential businesses. By identifying and isolating the contacts of infected people, Vietnam was able in nearly all cases to stop the spread.

Vietnam is admittedly an authoritarian state, where mandatory measures are more easily implemented than in a democracy, and observers might be tempted to question data offered by its government. In fact, the Vietnamese success has not only been affirmed by independent medical sources; it has been touted by media outlets on both the right and the left of the political spectrum.[13] Ironically, some observers in fact attribute the country's success in part to prompt, effective, and transparent information and communication campaigns to keep the public updated.[14]

While future work will be needed to analyze the Vietnamese experience, it is already clear that it has much in common with the experiences of China, Germany, Iceland, New Zealand, South Korea, and Taiwan. Political leaders in these countries took the threat seriously, attended to the advice offered by scientific experts, and established public health approaches based on that advice. They trusted science, and science repaid that trust by saving lives.

And of course it is not just COVID-19 that illustrates the importance of having and using scientific information. While the COVID-19 pandemic was unfolding, climate change continued to progress as well. The 2020 Atlantic hurricane season has been among the worst on record, with so many named hurricanes that we went through not only the entire Latin alphabet from A to Z, but the entire Greek alphabet as well.[15] Hurricanes are not just an inconvenience. They are not something to which people simply "adapt." They kill people, destroy homes, and, in the worst cases, leave permanent social, psychic, economic, and environmental damage. Meanwhile, while citizens of the Gulf

Coast were suffering a surfeit of rain, deadly wildfires were ravaging California and the Pacific Northwest.

Scientists have known for decades that climate change had the potential to make hurricanes and wildfires worse, and we have known for some years now that climate change *is* making these events worse. It has been many years since climate change was just a "theory." And yet, our political leaders continue to stall, prevaricate, and even deny outright the scientific realities. They listen not to the experts who have studied the problem and subjected their findings to the open criticism of fellow scientists, but to "anti-experts" who tell them not what is true, but what they want to hear.[16]

And so people get hurt. Their homes are destroyed. They die.

Not all of these deaths could be prevented by trusting science. We have, after all, always had hurricanes and pandemics and likely always will. Public policy will never be only a matter of listening to science, nor should it be. Many factors weigh into the decisions we make about our personal lives and our public policies, and rightly so. All choices are trade-offs; all public policies involve costs and benefits. But we cannot judge the trade-offs—we cannot accurately calculate the costs and the benefits—if we ignore (or worse, are deliberately denied) the relevant scientific information.

Put positively, a great deal of pain and suffering can be avoided when we understand scientific knowledge and put it to appropriate use. Scientists are people who understand things in ways that we can use to our advantage. They know things that we need to know. And, as COVID-19 has tragically proved, they know things that we ignore at our peril.

Notes

1. Talha Burki, "China's Successful Control of COVID-19," *The Lancet: Infectious Diseases* 20, no. 11 (October 8, 2020), https://doi.org/10.1016/S1473-3099(20)30800-8.

2. Pablo Gutiérrez and Seán Clarke, "Coronavirus World Map: Which Countries Have the Most COVID Cases and Deaths?," *The Guardian*, October 16, 2020, https://www.theguardian.com/world/2020/oct/16/coronavirus-world-map-which-countries-have-the-most-covid-cases-and-deaths; "COVID in the U.S.: Latest Map and Case Count," *The New York Times*, July 20, 2020, https://www.nytimes.com/interactive/2020/us/coronavirus-us-cases.html; Henrik Pettersson et al., "Tracking Coronavirus' Global Spread," CNN, accessed October 19, 2020, https://www.cnn.com/interactive/2020/health/coronavirus-maps-and-cases.

3. "Mortality Analyses," Johns Hopkins Coronavirus Resource Center, accessed October 19, 2020, https://coronavirus.jhu.edu/data/mortality.

4. Vietnam has a population over 97 million, but it has seen a tiny number of COVID-19 deaths: 38, as of October 19, 2020. Since the US population is 3.33 times as large, 3.33 × 38 = 127. "Vietnam COVID 19 Deaths—Google Search," accessed October 19, 2020, https://www.google.com/search?q=vietnam+COVID+19+deaths&oq=vietnam+COVID+19+&aqs=chromVietnam%20has%20a%20population%20of%2095%20million,%20but%20it%20has%20seen%20a%20tiny%20number%20of%20covid.0C19%20deaths:%2035e.1.0i457j0i20i263j69i57j0i20i263j0j69i60l3.5002j0j9&sourceid=chrome&ie=UTF-8.

5. "The Aging Readiness and Competitiveness Report: Germany," AARP, 2017, http://www.silvereco.org/en/wp-content/uploads/2017/12/ARC-Report-Germany.pdf. Cf. Deidre McPhillips, "Aging in America, in 5 Charts," *US News & World Report*, September 30, 2019, https://www.usnews.com/news/best-states/articles/2019-09-30/aging-in-america-in-5-charts.

6. "Pneumonia of Unknown Cause—China," WHO (World Health Organization), January 5, 2020, http://www.who.int/csr/don/05-january-2020-pneumonia-of-unkown-cause-china/en.

7. "Archived: WHO Timeline—COVID-19," WHO, accessed October 19, 2020, https://www.who.int/news/item/27-04-2020-who-timeline--covid-19.

8. Julia Naftulin, "WHO Says There Is No Need for Healthy People to Wear Face Masks, Days after the CDC Told All Americans to Cover Their Faces," Business Insider, accessed October 18, 2020, https://www.businessinsider.com/who-no-need-for-healthy-people-to-wear-face-masks-2020–4.

9. Ibid.

10. Since then, it has become clear that masks do work and perhaps even more effectively than some of their earlier advocates dared hope. See, for example, Stephanie Innes, "COVID-19 Cases in Arizona Dropped 75% after Mask Mandates Began,

Report Says," *The Arizona Republic,* accessed October 19, 2020, https://www
.azcentral.com/story/news/local/arizona-health/2020/10/09/covid-19-cases-az
-spiked-151-after-statewide-stay-home-order-and-dropped-75-following-local-mask
-man/5911813002. As I have argued elsewhere, it was scientifically logical that masks
should help at least somewhat; see Naomi Oreskes, "Scientists Failed to Use Common Sense Early in the Pandemic," *Scientific American,* November 2020, https://
www.scientificamerican.com/article/scientists-failed-to-use-common-sense-early
-in-the-pandemic.

11. "Vietnam COVID 19 Deaths—Google Search."

12. Todd Pollack et al., "Emerging COVID-19 Success Story: Vietnam's Commitment to Containment," *Our World in Data,* June 30, 2020, https://ourworldindata
.org/covid-exemplar-vietnam; Thi Phuong Thao Tran et al., "Rapid Response to the
COVID-19 Pandemic: Vietnam Government's Experience and Preliminary Success,"
Journal of Global Health 10, no. 2 (July 30, 2020), https://doi.org/10.7189/jogh.10
.020502020502.

13. George Black, "Vietnam May Have the Most Effective Response to COVID-19,"
The Nation, April 24, 2020, https://www.thenation.com/article/world/coronavirus
-vietnam-quarantine-mobilization; "How Did Vietnam Become Biggest Nation
without Coronavirus Deaths?," Voice of America, June 21, 2020, https://www
.voanews.com/covid-19-pandemic/how-did-vietnam-become-biggest-nation
-without-coronavirus-deaths.

14. Tran et al., "Rapid Response to the COVID-19 Pandemic."

15. National Oceanic and Atmospheric Administration, "2020 Atlantic Hurricane
Season," https://www.nhc.noaa.gov/data/tcr/index.php?season=2020&basin=atl.

16. On anti-experts, see Adapt by Sprout Social, "Combating Anti-Expert Senti-
ment on Social," May 8, 2018, https://sproutsocial.com/adapt/anti-expert
-sentiment.

A particularly damaging pandemic anti-expert is Scott Atlas, a radiologist and
senior fellow at the politically oriented, conservative Hoover Institution, who, with
no recognizable expertise in immunology, virology, epidemiology, or public health,
has been advising Donald Trump on COVID-19 in a manner that conflicts with the
views of most public health experts, including Drs. Anthony Fauci and Deborah Birx.
See Yasmeen Abutaleb and Josh Dawsey, "New Trump Pandemic Adviser Pushes
Controversial 'Herd Immunity' Strategy, Worrying Public Health Officials," *Washington Post,* August 31, 2020, https://www.washingtonpost.com/politics/trump
-coronavirus-scott-atlas-herd-immunity/2020/08/30/925e68fe-e93b-11ea-970a
-64c73a1c2392_story.html.

A related example of anti-experts muddying the intellectual waters around
COVID-19 involves the "Great Barrington Declaration," organized by the American
Institute for Economic Research; see "AIER Hosts Top Epidemiologists, Authors of

the Great Barrington Declaration," October 5, 2020, https://www.aier.org/article/aier-hosts-top-epidemiologists-authors-of-the-great-barrington-declaration. The American Institute for Economic Research is, as its name suggests, an economic institute with no recognizable claim to biological or medical expertise. Like many such institutes, it promotes a particular political agenda, in this case "free trade, individual freedom, and responsible governance." Those may or may not be good things, but they are not matters of science. AIER also promotes anti-scientific discussion of climate change, much of which promotes the familiar canard that climate change will be minor and manageable. One recent piece, for example, which discounted scientific interpretations of the dangers of climate-induced sea level rise was written not by a scientist but by a "writer, researcher, and editor on all things money, finance and financial history" (Joakim Book, "The Tide-Theory of Climate Change," October 28, 2020, https://www.aier.org/article/the-tide-theory-of-climate-change).

While pandemics do, of course, involve economic matters, the Great Barrington Declaration focused on the *public health response*, urging a herd immunity approach, which most public health experts consider to be a euphemism for allowing people to sicken and die. Indeed, expert estimates suggest that if the United States had undertaken that approach, more than 200 million people would likely have become ill, with the potential for more than 2 million deaths. This, of course, is comparable to the argument that we can just "adapt" to climate change. Of course we can, but at what price?

Moreover, the concept of herd immunity is normally invoked in the context of vaccination: What percentage of a population needs to be vaccinated in order to protect the whole population? In the absence of a vaccine, herd immunity typically means that at least 70% of a population will need to get sick before the population as a whole is protected. See Christie Aschwanden, "The False Promise of Herd Immunity for COVID-19," *Nature* 587, nos. 26–28 (October 21, 2020), https://www.nature.com/articles/d41586-020-02948-4; and Kristina Fiore, "The Cost of Herd Immunity in the U.S.," Medpage Today September 1, 2020, https://www.medpagetoday.com/infectiousdisease/covid19/88401.

The clearest argument against the herd immunity strategy is provided by a comparison of Sweden and Norway. According to a report in *Nature*, drawing on statistics from Johns Hopkins University, "Sweden has seen more than ten times the number of COVID-19 deaths per 100,000 people seen in neighbouring Norway (58.12 per 100,000, compared with 5.23 per 100,000 in Norway). Sweden's case fatality rate, which is based on the number of known infections, is also at least three times those of Norway and nearby Denmark"; see Aschwanden, "The False Promise of Herd Immunity for COVID-19." And the Swedish economy suffered, anyway, because the global economy is, well, global.

ACKNOWLEDGMENTS

This project would never have been completed without the considerable aid of my able and generous graduate student, Aaron van Neste, who helped me in countless ways. I am also deeply grateful to Erik Baker, Karim Bschir, Matthew Hoisch, Stephan Lewandowsky, Elisabeth Lloyd, Matthew Slater, Charlie Tyson, and an anonymous reviewer for comments on early drafts, and to all my students past and present, with whom I have thought through the question raised here. Whether Fleck was right about thought collectives, my own thinking has never been Cartesian.

Many of the ideas expressed here were developed over many years in the Science Studies Program at the University of California, San Diego (UCSD). I am grateful to UCSD colleagues past and present: Bill Bechtel, Craig Callender, Nancy Cartwright, Jerry Doppelt, Cathy Gere, Tal Golan, Philip Kitcher, Martha Lampland, Sandra Mitchell, Chandra Mukerji, Steven Shapin, Eric Watkins, and Robert Westman, with whom over many years I discussed the basis for scientific knowledge, truth, trust, proof, persuasion, and other weighty matters. I am also grateful to my current colleagues in the Department of the History of Science, Harvard University, with whom I have continued the conversation: particularly Allan Brandt, Janet Browne, Alex Cszisar, Peter Galison, and Sarah Richardson and to my

xx   •   Acknowledgments

colleagues in the "Assessing Assessment Project," Keynyn Brysse, Dale Jamieson, Michael Oppenheimer, Jessica O'Reilly, Matthew Shindell, Mark C. Vardy, and Milena Wazeck, who have helped me to explore and analyze what it is that scientists really do.

This project would not have been possible without the backing and enthusiasm of Stephen Macedo, Melissa Lane, and the Princeton University Tanner Lecture committee; Al Bertrand, Alison Kalett, and Kristin Zodrow at Princeton University Press; and the financial support of the Tanner Foundation. [I declare no competing financial interests.]

Above all, I am grateful to all the scientists, past and present, who have worked hard to earn our trust. I hope that in some small way this book in part repays that debt.

WHY
TRUST
SCIENCE
?

INTRODUCTION

Stephen Macedo

Science confronts a public crisis of trust. From the Oval Office in Washington and on news media around the world, the scientific consensus on climate change, the effectiveness of vaccines, and other important matters are routinely challenged and misrepresented. Doubts about science are sown by tobacco companies, the fossil fuels industry, free market think tanks, and other powerful organizations with economic interests and ideological commitments that run counter to scientific findings.[1]

Yet we know that scientists sometimes make mistakes, and that particular scientific findings now widely believed will turn out to be wrong. So why, when, and to what extent should we trust science?

These questions could hardly be more timely or important. As extreme weather events become more common, sea levels rise, and climate-induced migrations flow across borders, nations around the world confront mounting costs and humanitarian crises. Yet so-called experts do not always agree. A local television meteorologist may report that it is merely "some speculation from scientists" that global warming is contributing to extreme weather events, such as the "polar vortex" that hit the Upper Midwest and Northeast of the United States in late January 2019. On another channel, a scientist at a well-regarded research center insists that "we know why . . . . It's all because of human activities increasing the greenhouse gases in the atmosphere that trap a lot more heat down by the surface."[2]

As vitally important as climate science is to the future of humanity, that is only the tip of the iceberg. Are vaccines effective? Does the birth control pill cause depression? Is flossing good for your teeth? On these questions and so many others, scientists may agree yet doubts circulate. Who should we believe and why?

In *Why Trust Science?* Professor Naomi Oreskes provides clear and compelling answers to the questions of when and why scientific findings are reliable. She explains the basis for trust in science in highly readable prose, and illustrates her argument with vivid examples of science working as it should, and as it should not, on matters central to our lives. Readers will find here a vigorous defense of the trustworthiness of scientific consensus based not on any particular method or on the qualities of scientists, but on science's character as a collective enterprise.

A distinguished scientist and historian of science, Professor Naomi Oreskes has also emerged as one of the world's clearest and most influential voices on the role of science in society and the reality of man-made climate change.

This book grows out of the Princeton University Tanner Lectures on Human Values delivered by Professor Oreskes in late November 2016. On that occasion, four distinguished commentators, representing a variety of fields and perspectives, responded to Professor Oreskes's two lectures. This book contains the lectures, the four commentaries, and an extended reply by Professor Oreskes, all revised and expanded.[3]

Readers will find in the chapters that follow an overview of the leading philosophical debates concerning the nature of scientific understanding, scientific method, and the role of scientific communities. Oreskes defends the role of values in science, discusses the relationship between science and religion, and sets out her own credo as a scientist and defender of science. Our four commentators offer their perspectives on these issues, and

Oreskes closes with comments on the plight and promise of science in our time. A more detailed overview follows.

Why should we trust science? Professor Oreskes's initial answer is crisp and clear: scientific knowledge is "fundamentally consensual" and understanding science properly can help us "address the current crisis of trust."

Chapter 1 develops the problem of trust against the background of an account of philosophical debates about the nature of science and scientific method. In the eighteenth and nineteenth centuries, and before, trust often resided in "great men": science was regarded as trustworthy insofar as the scientists were. Gradually the alternative idea was advanced that careful observation and adherence to scientific methods were the bases of progress. Oreskes also surveys the varieties of empiricism that dominated philosophies of science in the first half of the twentieth century, and the challenge advanced by Karl Popper, who regarded the essence of science not as verification but openness to falsifiability, or "fallibilism."

Most important, on Oreskes's account, was the emergence of the idea of science as a collective enterprise. The "sociological view" of science was first advanced by Ludwik Fleck, in the 1930s, who held that the "truly isolated investigator is impossible . . . . Thinking is a collective activity." Oreskes endorses the idea that scientific progress depends on the collective institutions and practices of science, "such as peer-reviewed journals, and scientific societies through which scientists share data, grapple with criticisms, and adjust their views."

The central importance of scientific communities, their worldviews, and practices is the core of Professor Oreskes's view. When we focus on what scientists do, we find a variety of methods pursued with creativity and flexibility. She explores debates surrounding philosophies of science in the work of

Pierre Duhem, W.V.O. Quine, Thomas Kuhn, and others. She describes the social epistemology developed by feminist philosophers and historians of science, including the contributions of Helen Longino, who helped establish the idea that, as Oreskes puts it, "objectivity is maximized . . . when the community is sufficiently diverse that a broad range of views can be developed, heard and appropriately considered." Or, as she says later, "In Diversity There Is Epistemic Strength."

Professor Oreskes thus defends the "social turn" in our understanding of science while also describing the sense of threat that greeted the idea that scientific realities are socially constructed. Remember the obvious, she advises: scientists are engaged in sustained and careful study of the natural world. The empirical dimension is critical, but scientific expertise is also communally organized: objectivity arises from social practices of criticism and correction, most successfully in scientific communities that are diverse, "non-defensive," and self-critical.

We are warranted in placing "informed trust" in the "critically achieved consensus of the scientific community," argues Professor Oreskes. Individual scientists make mistakes, especially when "they stray outside their domains of expertise," and Oreskes provides some glaring examples. And science has no monopoly on insight into the natural world. Nevertheless, the practices and procedures of scientific communities increase the odds that scientific consensus is reliable.

We should trust the conclusions of the scientific community rather than the petroleum industry when it comes to climate change because the petroleum industry has a conflict of interest. It aims to profit by finding, developing, and selling petroleum resources, and it generally does that well. But those aims conflict with the pursuit of truth regarding climate change. As a general rule, we should be skeptical of the scientific claims of organizations guided by the profit motive or ones precommit-ted to an ideological point of view. Good science presupposes "that participants are interested in learning and have a shared in-terest in truth. It assumes that the participants do not have a major, intellectually compromising conflict of interest."

And yet, scientists sometimes get things wrong, so, Professor Oreskes asks in chapter 2, how do we know that they are not wrong now? If our knowledge is perishable and incomplete, how "can we warrant relying on it to make decisions, particularly when the issues at stake are often socially or politically sensitive, economically consequential, and deeply personal?"

To investigate these important questions, Oreskes examines five examples of science gone awry: what do these examples have in common, and what can we learn from them?

The first is the "Limited Energy Theory," popular in the late nineteenth century, which held that women should not partici-pate in higher education, on the grounds that energy expended on studying would adversely affect their fertility. The withering criticism to which this theory was subjected by Dr. Mary Put-nam Jacobi had, as the reader will learn, little immediate effect on male scientists.

Another example is the rejection of continental drift. Many American scientists in particular were hostile to the theory, which they argued was based on flawed "European" methodology.

A third example is eugenics, which is most closely associated nowadays with the Nazis, but which had a wide variety of advo-cates and practitioners in the United States and other Western countries. Oreskes provides a fascinating account of the complex politics of eugenics in the United States and Europe.

Oreskes's fourth example is hormonal birth control and the evidence that it often causes depression. Many women experi-ence the onset of depression after beginning certain birth control formulas, and Professor Oreskes relates her own experience. Yet medical science long discounted as unreliable the self-reports of millions of women.

Oreskes's final case is dental floss and the flurry of news reports asserting that there is no hard evidence that flossing is effective. Probing deeper, Oreskes argues that the lack of randomized trials to test for the effects of flossing hardly amounts to a lack of evidence.

From these diverse cases, Professor Oreskes draws some general lessons, which she groups under the themes of consensus, method, evidence, values, and humility.

The importance of hard-won scientific consensus, as an indicator of trustworthiness, holds up very well across the five cases. Oreskes also provides a fascinating discussion of the difficult question—*vital to the role of science in a democracy*—of non-expert opinion and how scientists should respond to it. Non-scientists—from nurses and midwives to farmers and fishermen—often have information or evidence relevant to science-based decisions. Patients have vital information about their symptoms. Yet, "Just because someone is close to an issue does not mean he or she understands it; conventional notions of objectivity assume distance for just this reason." The cases help illustrate and sharpen the distinction between reliable scientific authority and the interest and ideology-based pseudoscientific dissent we witness surrounding climate change, evolution, and vaccines.

Drawing from her five examples, Oreskes warns of the "methodological fetishism" that leads some scientists to dismiss valuable forms of evidence because they do not fit their methodological precommitments. Evidence comes in a variety of forms.

Values inevitably play a role in shaping science, Oreskes insists. In looking back on eugenics, scientists may say that science was distorted by values, but values were also central to opposing eugenics and also the Limited Energy Theory. Because values play an inevitable role, diverse scientific communities are more likely to be able to detect unexamined assumptions, blind spots, and inherited biases: "A community with diverse values is more likely to identify and challenge prejudicial beliefs embedded in, or masquerading as, scientific theory." She also allows that there can be legitimate non-scientific objections—including ones based on religious or moral values—to policies that are justified partly by science but also by particular value claims.

And humility is important. Diverse scientific communities can correct for the blind spots of arrogant scientists, but the history of science counsels humility: the greatest scientists (and, one might add, philosophers) have sometimes become fetishists about method, drawn false conclusions from evidence, and fallen prey to the prejudices and biases of their times.[4] Even the best of scientists should remember that a complete grasp of the whole truth is yet far beyond us.

So, when should we trust science? In concluding chapter 2, Oreskes summarizes: when an expert consensus emerges in a scientific community that is diverse and characterized by ample opportunities for peer review and openness to criticism. Of course, any particular scientific claim may be false, so she reminds us of Pascal's Wager: consider the stakes of error. It may not be certain that flossing will be good for your teeth, but it is cheap and easy. It may not be certain that human actions and policy changes can reverse the dire effects of climate change, but consider the calamities that await our children and grandchildren if we now ignore scientific predictions that are correct.

In a coda to her two lectures, Professor Oreskes returns to the issue of scientists' values. In theory, scientific findings are one thing and the question of what if anything to do about them is another. So one might suppose that whereas the practical question of "what is to be done" inevitably implicates values, the question of what scientific evidence shows need not. Ideally, science should be able to leave political and moral controversies to others.

Things are not so neat and simple, however. Professor Oreskes observes that people equate science with what they think are its implications. Fundamentalist and evangelical Christians from Williams Jennings Bryan to Rick Santorum have worried that evolutionary accounts of human origins undermine human dignity and morality, by making humans, in Santorum's words, "mistakes of nature." Skepticism about climate science, on the other hand, is fed by the suspicion that environmentalists seek to undermine the "American way of life": big cars, motorboats, and high consumption.

In the face of such suspicions it is profoundly mistaken, argues Oreskes, for scientists to retreat to value neutrality. In the face of the question: why should ordinary people trust science and take it seriously? It cannot be effective to reply that scientists lack values! That is precisely what worries people. Moreover, it is perfectly obvious that scientists do have values—everyone does—and that those values influence their work. To hide your values, Oreskes observes, is to hide your humanity.

So, scientists should be honest about their values. Many people will share those values, and on that basis trust can be built. The Creation revered by Christians is the biodiversity cherished by Scientists, says Oreskes, and the evidence is overwhelming that these are now gravely threatened.

In concluding, Professor Oreskes offers an eloquent summary of her own credo: her guiding values as a scientist and environmentalist. "If we fail to act on our scientific knowledge and it turns out to be right, people will suffer and the world will be diminished."

In the next section of this volume, four distinguished commentators expand upon, elaborate, or criticize central features of Professor Oreskes's lectures.

Professor Susan Lindee is the Janice and Julian Bers Professor of History and Sociology of Science at the University of Pennsylvania, where she also holds a variety of administrative posts. Lindee argues that in responding to scientific skepticism we should draw attention to the science that we encounter and rely upon constantly in our everyday lives. We should "work our way up, from the toaster," to the frozen peas, the smart phones, and the other miracles of modern science and technology that enhance our lives.

Of course, science's contributions are not always so positive. Professor Lindee reminds us of the twentieth century's brutal history of technology-enhanced warfare. She suggests that historians of science have sought to distance pure science from technological applications because of technology's profoundly mixed legacy. Atomic scientists sought to maintain their moral purity by attributing the design of the bomb to mere engineers.

Marc Lange is the Theda Perdue Distinguished Professor and department chair in philosophy at the University of North Carolina, where he specializes in the philosophy of science. Lange notes that the question of why we should trust science seems to lead into a vicious circularity: isn't peer review just experts vouching for other experts?

Professor Lange suggests that asking for an external vindication of science as a whole may be unreasonable: science is self-correcting in that it can subject any particular scientific claim to critical scrutiny, "But science *cannot* reasonably be expected to put *all* its theories in jeopardy *at once.*"

Lange also raises the issue of what Thomas Kuhn described as revolutionary challenges to entire worldviews or paradigms, in which methods and theories "interpenetrate." Using the example of Galileo, he suggests that there is typically "sparse common ground" across paradigm shifts, and scientists can use it to build an argument for one of the rival theories against the others. Lange closes by urging philosophers and others to stop overemphasizing "incommensurability and under-determination" and to devote more attention to positive accounts "of the logic underlying scientific reasoning."

Ottmar Edenhofer is deputy director and chief economist at the Potsdam Institute for Climate Impact Research, as well as a professor at the Technical University Berlin. He offered a comment in Princeton, and is joined here by Martin Kowarsch, who is head of the working group on Scientific Assessments, Ethics, and Public Policy at the Mercator Research Institute. They begin by suggesting that the Trump administration accepts much climate science but opposes ambitious climate change mitigation efforts, partly because it heavily discounts the costs of climate change outside the United States. Thus, scientific consensus does not equal policy consensus, and so they ask how Oreskes's account of trust in science may need to be extended or amended for science-based policy assessments. They advise experimentation aimed at incremental learning about alternative policy pathways, and argue that costly mistakes have been made due to insufficient awareness of the complexity of the policy alternatives.

Edenhoffer and Kowarsch agree with Oreskes that value neutrality is impossible. They build on Deweyan pragmatism to propose that all socially important values—"equality, liberty, purity, nationalism, etc."—should be included in policy assessments: this may open the door to new and creative proposals.

Finally, Jon Krosnick offers some thoughts, inspired by Professor Oreskes's lectures, on the current state and future of science. Krosnick is Frederick O. Glover Professor in Humanities and Social Sciences and professor of communication, political science, and psychology at Stanford University, where he also directs the Political Psychology Research Group.

Professor Krosnick describes a number of famous (now infamous) and influential scientific findings—in biomedicine, psychology, and elsewhere—whose results scientists have been unable to replicate. In some cases the data were fabricated, in other cases investigators admitted to repeating an experiment until the desired result was produced.

Flawed research results partly from faulty methods, argues Krosnick, and also the desire for career advancement. Academic departments and professions place a premium on publishing surprising and counterintuitive findings. Is it any wonder that many of these prove unfounded on closer inspection? Journals rarely publish negative results so refutation of bad research is slowed. He insists that scientists must face up to the problems and address the counterproductive motivations that are now rampant.

In her wide-ranging *Reply to Critics*, Professor Oreskes deepens and enriches her argument.

She praises Susan Lindee for her brilliant historical account of scientists' attempts to distance themselves from the technological applications of their work, yet expresses doubt that becoming clearer-eyed about the science embodied in frozen peas and smart phones will have much effect on people's attitudes to climate science. Americans do not reject science in general but rather particular "scientific claims and conclusions that clash with their economic interests or cherished beliefs."

In response to Marc Lange, Professor Oreskes expresses doubt that trust in scientific experts is viciously circular. The "social markers of expertise are evident to non-experts," she argues, and it is relatively easy to figure out that climate science deniers are non-experts and that the American Enterprise Institute is pre-committed to certain policy outcomes. Expert scientific consensus does tend to be reliable.

In response to Edenhofer and Kowarsch, Professor Oreskes agrees that more work is needed on how to move from science to policy. Yet she insists that when powerful actors to seek to undermine public trust in the science associated with progressive climate policy, the *roots* of their skepticism are typically not in distrust of science but rather in economic self-interest and ideological commitments. Oreskes reiterates that if scientists are honest about their values, as she recommends, then they will often find that there is considerable overlap on the values behind climate policy disagreements, and this may help us build greater trust.

Professor Oreskes turns, finally, to Jon Krosnick's assertion that science faces a "replication crisis." While allowing that there have been notable examples, often involving the misuse of statistics, she points out that the rate of retractions—that is, retractions as a percentage of published articles—is tiny: perhaps less than .01%. If the rate has risen, that may reflect a salutary increase in critical scrutiny of findings, rather than a higher incidence of faulty research. Or it may reflect unwarranted media coverage of flashy single-paper results in psychology and biomedicine.

Oreskes pushes back against Krosnick's wider suggestions about a crisis in science. His examples furnish no evidence that fraud is commoner in science than elsewhere. Moreover, in some of Krosnick's examples fraud was discovered and punished expeditiously. Refutation and retraction are paths to progress. She reminds us that her argument has been that we should trust scientific *consensus*, not the single studies to which Krosnick draws attention, and reiterates that motivated industry funding of research is a serious problem.

In an afterword penned just before this book went to press, Professor Oreskes notes that the problem of trust in science—and in news and information more generally—has exploded since she delivered the Princeton Tanner Lectures in the fall of 2016. Many more Americans believe in the reality of climate change than once did, but America is led by a science and fact-denying chief executive who is reversing hard-won progress on climate policy. It remains the case that much doubt about consensus findings in science is manufactured by those with financial or ideological interests in derailing science-based policies, just as she and Erik Conway argued in *Merchants of Doubt*.

Professor Oreskes closes by reiterating that science merits our trust when scientific results achieve consensus among the expert members of diverse and self-critical scientific communities. And she offers a final example—controversies over the use of sunscreen—to illustrate this book's core theme.

Like all excellent books, this one addresses many questions and also raises some. While Professor Oreskes argues that progress and reliability in science depends more on the qualities of scientific communities than on the character of individual scientists, she also argues that scientists' inevitably have values and that they should be honest about them. Do not well-working scientific communities depend on the predominance of good values—of intellectual honesty and truth seeking—among scientists? And if diversity is important in scientific communities, of what kinds? The inclusion of women and members of racial, ethnic, religious, and other minority populations has obviously been very good for all of the sciences, and scholarship generally. Are there social sciences (and perhaps other fields of inquiry) in which greater ideological diversity would be helpful?

Readers will come away from this volume armed with a far better understanding of the vitally important enterprise of modern science and the reasons why we should trust scientific consensus. All who care about the future of humanity on this fragile earth should hope that this timely and important book gains a wide audience, before it is too late.

Chapter 1

WHY TRUST SCIENCE?

Perspectives from the History and Philosophy of Science

The Problem[1]

Many people are confused about the risks involved in vaccination, the causes of climate change, what to do to stay healthy, and other matters that fall within the domain of science. Immunologists tell us that vaccines are generally safe for most people, have protected millions of people from deadly and disfiguring diseases, and do not cause autism. Atmospheric physicists tell us that the build-up of greenhouse gases in the atmosphere is warming the planet, driving sea level rise and extreme weather events. Dentists tell us to floss our teeth. But how do they know these things? How do we know they're not wrong? Each of these claims is disputed in the popular press and on the internet, sometimes by people who claim to be scientists. Can we make sense of competing claims?

Consider three recent examples.

One: In a 2016 presidential debate, Donald Trump rejected the position of medical professionals—including that of fellow candidate physician Ben Carson—on the safety of vaccination. Recounting the experience of an employee whose child was vaccinated and later diagnosed as autistic, Mr. Trump stated his view that vaccines should be given at lower doses and be more widely spaced. Few medical professionals share his view.[2] They consider delaying vaccination to increase the risk that infants and children will contract dangerous and otherwise preventable diseases such as measles, mumps, diphtheria, tetanus, and pertussis. Some of the children who contract these diseases will become gravely ill or die. Others will survive but pass on the infections to others. Yet, Mr. Trump is not alone in making this suggestion; prominent celebrities have made similar exhortations. Many parents now reject the advice of their physicians and choose to have their children vaccinated on a delayed schedule—or not at all. As a result, morbidity and mortality from preventable infectious diseases are on the rise.[3]

Two: The vice president of the United States, Mike Pence, is a young Earth creationist, meaning that he believes that God created the Earth and all it contains less than ten thousand years ago. The consensus of scientific opinion is that Earth is 4.5 billion years old, that the genus *Homo* emerged two to three million years ago, and that anatomically modern humans appeared about two hundred thousand years ago. While science cannot answer the question of whether God (or any supernatural being or force) guided the process, most scientists are persuaded that life on Earth evolved largely through the process of natural selection over the course of Earth's history, that humans share a common ancestor with chimpanzees and other primates, and that divine intervention is not required to explain the existence of *Homo sapiens sapiens*.[4]

Do Americans lean toward the scientific view or the Pencian view? The answer depends a bit on how you ask the question, but if you are a religious person in America who attends church regularly, the chances are high that you agree with Mike Pence: 67% of regular churchgoers believe that God created humans in their present form within the last ten thousand years. Some of us may think that these people are all Republicans, but we would be wrong. According to the Gallup polling organization, while 58% of Republicans agreed with the statement that "God created humans in their present form, within the last 10,000 years," so did 39% of independents and 41% of Democrats.[5] Given this popular support for creationism, it is perhaps unsurprising that in 2012, the state of Tennessee enacted what some have called a "twenty-first-century Monkey Law," empowering teachers to teach creationism in science classrooms.[6] Despite repeated rejection of previous laws of this type by US courts, many states continue to attempt to enact comparable laws.[7]

Three: The American Enterprise Institute (AEI) is a long-established and well-funded think tank in Washington, DC, committed to principles of laissez-faire economics, market-based mechanisms to social problems, limited (federal) government, and low rates of taxation. The Institute has long promoted skepticism about the scientific evidence for anthropogenic climate change and disparaged the conclusions of the scientific community, including the Intergovernmental Panel on Climate Change (IPCC).[8] AEI scholars have suggested that climate scientists are suppressing dissent within their community; the Institute at one point offered a cash incentive to anyone willing to search for errors in IPCC reports. Jeffrey Sachs, head of the Earth Institute at Columbia University from 2002–16 and special advisor to UN secretary-general António Guterres on the Millennium Development Goals, has said of one well-known AEI scholar that he "distorts, misrepresents, or simply ignores" relevant scientific conclusions.[9] In 2016, this particular scholar referred to scientists as an "interest group," demanding to know why "scientific analysis conducted or funded by an agency headed by political appointees buffeted by political pressures . . . [should] be viewed ex ante as any more authoritative than that originating from, say, the petroleum industry?"[10]

I am no fan of the American Enterprise Institute. With my colleague Erik M. Conway I have shown how they (along with other think tanks promoting laissez-faire approaches to social and economic issues) have persistently mispresented or mischaracterized scientific findings on climate change, as well as a variety of public health and environmental questions. (They are no fans of mine, either. Their scholars have attacked my work on scientific consensus.)[11] But the question raised is a legitimate one. Should a scientific analysis be viewed as ex ante authoritative? Is it reasonable to take the default position that the scientific community can in general be trusted on scientific matters, but the petroleum industry (to use his example) cannot?

Science in North American universities and research institutes is generally well funded and respected—typically much more so than the arts and humanities—but outside those hallowed halls something very different is transpiring. The idea that science should be our dominant source of authority about empirical matters—about matters of fact—is one that has prevailed in Western countries since the Enlightenment, but it can no longer be sustained without an argument.[12] *Should* we trust science? If so, on what grounds and to what extent? What is the appropriate basis for trust in science, if any?

This is an academic problem but one with serious social consequences. If we cannot answer the question of why we should trust science—or even if we should trust it at all—then we stand little chance of convincing our fellow citizens, much less our political leaders, that they should get their children vaccinated, floss their teeth, and act to prevent climate change.

Scholars' views on the answer to this question have changed dramatically and more than once in the past century. Moreover, some of the answers that scientists offer are manifestly contradicted by historical evidence. It is routine, for example, for scientists to insist that their theories must be correct, because they *work*. How else, they argue, would planes fly or medicines cure disease?[13] But utility is not truth: we can identify many theories in the history of science that worked and later were rejected as wrong. The Ptolemaic system of astronomy, the caloric theory of heat, classical mechanics, and the contraction theory of the Earth explained observed phenomena and made successful predictions, and are now on the scrap heap of history. Many scholars in the history and philosophy of science and science studies have, however, recently converged on a new view that does hold up to scrutiny: of scientific knowledge as fundamentally *consensual*. This consensual view of science can help us address the current crisis of trust.

The Dream of Positive Knowledge

Throughout the eighteenth and the early nineteenth centuries, most scholars located the authority of science in the authority of the "man of science."[14] The results of scientific investigations were trustworthy to the extent that the people who undertook them were. This is one reason why scientific honor societies, such as the Royal Society or the Académie des Sciences, were created: to acknowledge and identify the "worthies" whose opinions on scientific matters should be sought, trusted, and heeded.[15] These societies served to identify the individuals whose work was considered worthy of acceptance. In the United States, this ideal was instantiated in the creation of the US National Academy of Sciences during the Civil War to advise President Lincoln. Identifying these "great men" of science would enable the president to get the reliable advice he needed.

However, in the mid-nineteenth century, a substantive intellectual shift occurred, driven to a significant extent by the work of Auguste Comte (1798–1857), variously credited as the founder of sociology, the founder of philosophy of science in its modern form, and the founder of the philosophical school known as positivism.[16] Comte's work is abundant and complex and has been subject to various considerations and reconsiderations, refutations and restorations, but the most important aspect, for our purposes, is his commitment to the idea of *positive knowledge*. Science, Comte believed, was uniquely able to provide positive—which is to say *reliable*—knowledge. While the term "positive knowledge" is no longer much used apart from academics discussing it, most often as a discredited concept, the idea persists in our linguistic conventions. We still retain the notion of something being "absolutely, positively true." In English we can ask: "Are you positive?" To which you may reply: "Yes, I'm positive."

For Comte, the key element in the concept of positive knowledge is *method*, which he contrasted with *doctrine*—whether religious, superstitious, or metaphysical. The doctrines of religion and metaphysics, he argued, were forms of bias and blinkering that impeded intellectual and social progress, which the method of science, by contrast, could provide. By applying method to the pursuit of knowledge, science had the potential to liberate men and women from the shackles of religion and superstition.

Comte's philosophy (like many in the nineteenth century, including famously Marxism) was teleological: he saw human history as being characterized by three stages: the theological or fictitious, the metaphysical or abstract, and the scientific or positive. These were not necessarily sequential—they might coexist within a society or even with a person—but overall the direction of progress was from theology to science, with metaphysics serving as a necessary transition.[17] In the "positive stage" of human development, theology and metaphysics are replaced by scientific reasoning. And scientific reasoning is rooted in observation.

It has been argued that Comte was seeking to replace conventional religion with a new religion of science, and there is some justice to this claim. Teleology is a common feature of many religions. He accepted that people had a need for moral principles but thought those principles could be found in the humanistic ideals of truth, beauty, goodness, and commitment to others. He also believed that people had a need for ritual and proposed to replace the veneration of Christian saints with a set of positivist heroes. In his own life, he set aside time for meditation and affirmation of his central values.[18] But whether his views were quasi-religious or not, the key point for our discussion is that for Comte—and generations of those who followed him, knowingly or not—science was reliable because of its commitment to method. This leads one to ask: what is that method?

Comte was sensitive to the variety of scientific disciplines that were developing at that time. He did not assert that their practices were uniform, but he did believe that they shared a fundamental characteristic of the "positive" state of human existence. He wrote:

In the positive state, the human mind, recognizing the impossibility of obtaining absolute truth, gives up the search after the origin and hidden causes of the universe and a knowledge of final causes of phenomena. It endeavours now only to discover, by a well-combined use of reasoning and observation, the actual laws of phenomena—that is to say their invariable relations of succession and likeness. The explanation of facts, thus reduced to its real terms, consists henceforth only in the connection established between different particular phenomena and some general facts, the number of which the progress of science tends more and more to diminish.[19]

In stressing the importance of empirical regularities, Comte was making an argument similar to the British empiricists, particularly David Hume.[20] He acknowledged his debt to British empiricism, particularly the work of Francis Bacon, writing, "All competent thinkers agree with Bacon that there can be no real knowledge except that which rests upon observed facts."[21] But he was not the "naïve positivist" that some later commentators made him out to be. He was a sophisticated thinker who recognized that our theories structure our observations as much as our observations structure our theories:

> If we consider the origin of our knowledge, it is no less certain that . . . [as] every positive theory must necessarily be founded upon observations, it is, on the other hand, no less true that, in order to observe, our mind has need of some theory or others. If in contemplating phenomena we did not immediately connect them with some principles, not only would it be impossible for us to combine these isolated observations and, therefore, to derive any profit from them, but we should even be entirely incapable of remembering the facts, which would for the most part remain unnoted by us.[22]

We can understand, therefore, why primitive humans had need of religion, superstition, and metaphysics: these early concepts were a step toward apprehending the world around us. We need not disdain or disparage these early stages in human development, we simply need to recognize and accept that to move forward—to identify the true laws that govern nature—our thinking needs to be grounded upon observation. In his words: "we must proceed sometimes from facts to principles [and] at other times from principles to facts," but ultimately we will establish "as a logical thesis that all our knowledge must be founded upon observation."[23]

Comte was also a fallibilist: he recognized that our views would grow and change and that his own vision would in time be modified. (Indeed, if his basic concept was correct, then the progress of knowledge would necessarily modify our views, and we might note that the persistence of religion has falsified a key element of his teleology.) But, to his credit, Comte was consistent insofar as he insisted that future change in our thinking would be the outcome of our observations.

Comte was also reflexive, recognizing that the practices of observation must themselves be subject to observation. An improved knowledge of positive method must come, therefore, not by *theorizing* it but by studying it; we must observe science in order to understand it. Comte thus anticipated Bruno Latour and his anthropological studies of laboratory science by more than a century when he held: "When we want not only to know what the positive method consists in, but also to have such a clear and deep knowledge of it to be able to use it effectively, we must consider it *in action*."[24]

Comte's key move was to insist that science is reliable not by virtue of the character of its practitioner, but by virtue of the nature of its practices.[25] We need to attend to these practices by studying them empirically. The key questions, then, for those who took up the Comtean program were: What exactly are those practices? Is there a scientific method?

Varieties of Empiricism

For twentieth-century empiricists, which we have come to call logical positivists or logical empiricists, the answer to the question of the method of science was the principle of verification.[26] The concept was developed most extensively by a group of German-speaking philosophers and scientists, known as the "Vienna Circle." The most famous English language articulation of the verificationist program came from the Oxford philosopher A. J. Ayer (1910–89). In his 1936 book, *Language, Truth and Logic,* which is still in print, Ayer summarized the principle by framing it in terms of the problem of meaning: A statement can be considered meaningful if and only if it can be verified by reference to observation. Put another way, "some possible observation must be relevant to the determination of [the statement's] truth or falsehood."[27] Science is the practice of formulating meaningful statements, and using observations to judge whether a meaningful statement is correct.

Verification gives us the basis for evaluating what is or is not justified true belief. If a claim can be verified through observation, and if it has in fact been so verified, then we are justified in believing it, which is to say, justified in accepting it as true. If a claim cannot be so verified, then it is meaningless and need not detain us further. Thus, in one fell swoop did Ayer dispense with religion, superstition, and various forms of political ideology and theory that were unverifiable. The principle of verification provided a means of demarcating scientific knowledge from non-scientific knowledge: scientific claims were verifiable thorough observation; claims that were not verifiable were not scientific.

Like Comte, Ayer was ambitious but not naïve. He understood that in practice any observation necessarily entails background assumptions. But, like his Vienna Circle colleagues Rudolf Carnap and Otto Neurath, he insisted that verification through observation was the key component to meaning, hence the moniker *verificationism*. In order to test a statement, one had to be able to deduce an observable consequence from it and express that deduction as a *statement*, and that deduction had to be specific to the statement under investigation for the verification to be dispositive. Ayer wrote: "A statement is verifiable, and consequently meaningful, if some observation statement can be deduced from it in conjunction with certain other premises, without being deducible from those other premises alone."[28]

Ayer and his colleagues recognized that any program that foregrounded observation necessarily faced the problem of induction: namely, how many observations are needed to conclude that a statement is true? Following Hume, his answer was that inductive knowledge was necessarily probabilistic, and he suggested that one needed to allow for weak and strong forms of verification, based on the quantity and quality of available relevant observations. These sorts of concerns underpinned research on the character of scientific observation, which quickly led to various complications regarding the formulation of observation statements, the meaning of terms, and the identification of what, precisely, was being verified by any particular observation or set of observations.

These issues detained many logical empiricists for the rest of their lives. Carl Hempel, in particular, paid attention to the role of hypothesis in generating testable observation statements; Carnap focused on the observation statements and the language in which they were rendered, and famously argued with Willard Van Orman Quine over whether observations could really confirm or refute beliefs. (Quine concluded they could not, a point we will take up.) This work did not resolve the issues it entailed.[29] For our purposes, the important point is that the logical empiricists sustained the central Comtean idea that the core of scientific method is verification through experience, observation, and experiment.

Challenges to Empiricism

While logical empiricism is often attacked as the ruling dogma of twentieth-century philosophy of science, it never really ruled. Even in its heyday, several important challenges were already underway.[30]

Karl Popper and Critical Rationalism

The most well-known critic of logical empiricism is Karl Popper (1902–94). Popper rejected several key tenets of logical positivism. First, he denied that induction was the method of science. Second, he argued that what distinguishes science from other forms of human activity is not its activities, but its attitude. Great scientists are notable for the critical attitude they take toward their work, which is an attitude of skepticism and disbelief. Third, he insisted that the goal of science is not to prove theories—since that cannot be done—but to disprove them. He introduced his now-famous notion of *falsifiability*, concluding that what distinguishes a scientific claim from a non-scientific one is not that there is some observation by which it can be verified, but that there is some observation by which it can be refuted.

These three ideas are linked in the following way. There may be habits or practices or even principles of induction, but there is no rational rule of induction. Inductive inferences cannot be justified based on any purely logical rule, and therefore cannot be established with logical necessity. This is what nowadays is referred to as the black swan problem. I may have observed one hundred swans, or one thousand, or ten thousand, and found that they have all been white, as have all the swans observed by my scientific colleagues. Therefore, my colleagues and I conclude (seemingly with robust warrant) that all swans are white. Yet, one day I travel to Perth, Australia, where I see a black swan.

Thus, we see that observations cannot prove that a theory is true, no matter how extensive or comprehensive. Refutation may be lurking around the corner (or the antipodes). If science is to be a rational enterprise, induction therefore cannot be its method.

Because observation alone cannot give us *logical* grounds to support inductive generalizations, verification cannot be the basis of scientific method. However, the observation of the black swan did prove that my inductive generalization was false, so there is a logic of *refutation*. There is a logical asymmetry between verification and falsification: verifications are always necessarily provisional, whereas falsifications (Popper held) can be dispositive. Given this, as a scientist I should not be seeking observations that confirm my theory, but observations that might refute it. The method of science, Popper therefore concludes, is neither generalization from observation nor verification by observation, but *falsification*. Put another way, the key activity of science is not the gathering of observations, but the formulation of conjectures and the pursuit of specific observations that may refute them. Thus the title of his famous collection of essays and lectures: *Conjectures and Refutations*.

Even more urgently than his logical positivist colleagues, Popper held science to be the model of rationality, insisting that critical rationality is not only the appropriate basis for intellectual inquiry, but also for politics and civil society, as it empowers resistance to authoritarianism of both the right and the left. Therefore he labeled his approach *critical rationalism*. His project was both epistemological and political: he sought an epistemology that would enable not just scientific rationality but also political rationality in democratic forms of governance. Among other things, Popper sought to refute Marxism by showing that "scientific socialism" was an oxymoron, because problems in Marxist theory were never taken as refutations but only as elements to be explained or accounted for in some way.[31]

Popper's critical rationality ironically opened the door for a form of radical skepticism that he abhorred. Popper pushed fallibility further than his predecessors, insofar as he insisted that refutation is not merely an inevitable feature of science, but the goal of it; it is through refutation that science advances. But if our scientific views are not only soon to be refuted, but *should* be refuted, then why should we believe any of it?[32] Popper's answer was to develop the notion of corroboration: that we can have good reason to believe theories that have passed severe tests, such as the deflection of starlight as a test of the general theory of relativity. Successful empirical tests corroborate theories, even if they do not prove them. In making this move, Popper helped to explain why theory testing plays such a major role in scientific practice, but he also radically weakened the otherwise strict tenor of his work: we are now left with having to make subjective judgments as to what constitutes a "severe" test and how many such tests we need.

Ludwik Fleck and Thought Collectives

The various forms of positivism that developed from the mid-nineteenth to the mid-twentieth century were all concerned with method, paying less attention to the people who were pursuing that method or the institutional structures within which they operated. Popper paid some heed to the character of the individual scientist, insofar as he stressed the importance of a critical investigative attitude. But Popper's epistemology (like his political theory) was individualistic; he vested the advance of science in the actions of the bold individual who doubted an existing claim and found a means to refute it. Popper paid less attention to the institutions of science, and was actively hostile to suggestions of collectivism, redolent as they were of the Marxist philosophy and Communist politics that he loathed.[33]

The recognition of science as a collective activity thus laid the grounds for a radical challenge to received views of science that would flourish in the second half of the twentieth century. Whether one had read Comte or Ayer or Popper, one could have come away with the impression that scientists, like Descartes in his room staring at melting wax, lived, worked, and thought alone. Yet anyone who studied science in action—as Comte instructed us to do—or who participated in scientific research knew that wasn't so. Yet somehow this had escaped sustained scholarly attention.

Ludwik Fleck (1896–1961) changed that. A microbiologist who made the social interactions of scientific life a centerpiece of analysis, in hindsight he is credited with developing the first modern sociological account of scientific method. In his 1935 work, *The Genesis and Development of a Scientific Fact: An Introduction to the Theory of Thought Style and Thought Collective*, Fleck shifted attention from the individual scientist to the activities of communities of scientists, and proposed that scientific facts are the collective accomplishment of communities. In doing so, he pioneered the analysis of the social interactions that yield scientific facts.

Fleck was aware of the logical positivists' work; he sent his work to the Viennese positivist Moritz Schlick seeking help to get it published.[34] He was also in contact with historians and philosophers of medicine and mathematics in Poland at that time. But scholars have mostly concluded that his work was primarily influenced by his experience as a researcher and his attention to developments in science, particularly the rise of quantum mechanics in physics, which (he believed) had led to the emergence of new styles of thinking.

Fleck's key point was that scientists worked in communities in which styles of thought became shared resources for future work, including the interpretation of observations. He labeled these communities "thought collectives." Groups of scientists within any particular discipline—biology, physics, geology—constituted thought collectives whose common ways of thinking made it possible for them to work together, share information, and interpret that information in meaningful ways. Without a thought collective, science could not exist. He wrote:

> A truly isolated investigator is impossible . . . Thinking is a collective activity. . . . Its product is a certain picture, which is visible only to anybody who takes part in this social activity, or a thought which is also clear to the members of the collective only. What we do think and how we do see depends on the thought-collective to which we belong.[35]

The term "thought collective" may invoke the specter of thought police, and Fleck recognized that collectives could be conservative or even reactionary—as he believed religious thought collectives were. But a thought collective could also be democratic and progressive, and this was the key to understanding science. Science (unlike most European religion) has a democratic character: all researchers can participate in an equitable way, and through their interactions with each other, refine and change the views of the whole.

Fleck had a radical view of how far such change could go, stressing that over time changes could be so great that the meanings of terms changed, that problems that were previously seen as central could now be dismissed as irrelevant or even illusory, and new issues would emerge that previously went unrecognized. While the increments of change were small—the pathways of change more evolutionary than revolutionary—eventually the thought style may have changed so much that the old view is essentially unrecognizable, even indecipherable.

> Thoughts pass from one individual to another, each time a little transformed, for each individual can attach to them somewhat different associations. Strictly speaking, the receiver never understands the thought exactly in the way that the transmitter intended it to be understood. After a series of such encounters, practically nothing is left of the original content.[36]

Scientific ideas, like evolution itself, may change dramatically over time, but they do so by the accumulation of small transformations and differing interpretations.

"Whose thought is it that continues to circulate?" Fleck asks. His answer: "It is one that obviously belongs not to any single individual but to the collective."[37] As Helen Longino would later put it in a slightly different context, "Of course, Galileo and Newton and Darwin and Einstein were individuals of extraordinary intellect, but what made their brilliant ideas *knowledge* were the processes of critical reception." Fleck would say: of reception and transformation.[38] Newtonian mechanics is not equivalent to the contents of the *Principia*, nor is evolutionary biology coincident with the contents of the *Origin of Species*. The ultimate outcome is the result of Newton and Darwin's work and the diverse ways in which over time it has been interpreted, adjusted, and altered.

Scientific progress in this view is inextricably connected with the institutions of science such as conferences and workshops, books and peer-reviewed journals, and scientific societies through which scientists share data, assess evidence, grapple with criticisms, and adjust their views. Scientific research is organized, it is cooperative and interactive, it creates shared worldviews, and observations are interpreted in accordance with these worldviews. Progress, Fleck holds, consists of the revision and adjustment of worldviews as the community deems appropriate, and over time these adjustments may be so great as to constitute a new worldview, a new style of thought, even a new reality.[39] What the thought collective previously recognized as physical reality may no longer be viewed as reality. Fleck is unambiguously anti-realist on this point: what members of a collective call truth is merely what the thought collective has settled upon at that point. He is also unambiguously anti-individualist and anti-methodological: the agency of scientific progress is located not in the individual but in the group, and the core of science lies not in a particular method but in the diverse interactions of that group.

Under-determination: Pierre Duhem

Fleck's work received some attention when first published, but became much more famous in later years when it came to be viewed as anticipating and influencing the work of Thomas Kuhn. Something similar may be said about Pierre Duhem (1861–1916), whose work was recognized by the Vienna Circle but is now seen as influential primarily because of its uptake by the American philosopher W.V.O. Quine (1908–2000).

To scientists, Duhem is known as a founder of chemical thermodynamics, but he was also a sedulous historian and acute philosopher of science.[40] To philosophers and historians of science today, he is best known for his 1906 book, *The Aim and Structure of Physical Theory*, with its refutation of the notion of a critical experiment and its articulation of what has come to be known as the principle of under-determination.[41]

Duhem's central argument was simple: The Baconian idea of a crucial experiment is mistaken, because if an experiment fails there are many reasons why that might be, so we don't necessarily know what has gone wrong. Conversely, if an experimental test of a theory succeeds, other consequences of the theory may yet be shown to be incorrect. The support for a theory must in principle include all the potential tests of it, and its refutation must be considered in light of all the possible elements that were necessary to perform the experiment in the first place. As the physicist Louis de Broglie put it in 1953 in the preface to the English edition:

> According to Duhem, there are no genuine crucial experiments because it is the ensemble of a theory forming an individual whole which has to be compared to experiment. The experimental confirmation of one of its consequences, even when selected among the most characteristic ones, cannot bring a crucial proof to theory, for . . . nothing permits us to assert that other consequences of the theory will not yet be contradicted by experiment, or that another theory yet to be discovered will not be able to interpret as well as the preceding one the observed facts.[42]

Put simply: any test of a hypothesis is simultaneously a test of the specific hypothesis under consideration and of the experimental setup, auxiliary hypotheses, and background assumptions. A failed experiment does not necessarily reveal where the failure lies, and a successful experiment does not preclude that a different experimental arrangement or other auxiliary hypotheses would have revealed some difficulty. Duhem wrote: "Any experimental test [in physics] puts into play the most diverse parts of physics and appeals to innumerable hypotheses; it never tests a given hypothesis by isolating it from the others."[43]

Nor does experimental evidence exhaust the range of possible theoretical options open to us: Duhem was explicit that hypotheses are not simply inductions from observation. It is impossible, he asserted without equivocation, to "construct a theory by a purely inductive method."[44] Both theory and experiment have a role in science, and it is mistaken to view experiments as more crucial than theory, mistaken to view them as the source of theory, and above all, mistaken to view them as the final arbiter of theory.

Duhem was not rejecting experimentation. On the contrary, he argued that "the sole purpose of physical theory is to provide a representation and classification of experimental laws."[45] Experiment is essential both to discovering those laws in the first place and to testing the general physical theories that we develop to account for them. The "only test permitting us to judge a physical theory and pronounce it good or bad is the comparison between the consequences of this theory and the experimental laws it has to represent and classify." This view is essentially probabilistic: an experiment can neither verify nor refute a theory; rather it simply tells us whether a theory is "confirmed or weakened by the facts."[46]

De Broglie suggested that a key to Duhem's thought was his interpretation of Léon Foucault's famous experiment in which he demonstrated that the speed of light in water is less than its speed in a vacuum, taken by many as a crucial experiment validating the wave (as opposed to particle) theory of light. Duhem disagreed. Even if Foucault's experiment contradicted Newton's corpuscular theory, other forms of corpuscular theory might yet be consistent with the result.[47]

Yet Duhem did not adopt the radical holism with which his name later became associated. (Holism is the idea that theories stand or fall in their entirety and that a challenge to any one component is potentially a challenge to the entire intellectual fabric.) In places, it may appear that he is on the verge of radical holism, as when he writes of the "radical impossibility [of separating] physical theories from the experimental procedures appropriate for testing these theories," or that an "experiment in physics can never condemn an isolated hypothesis but only a whole theoretical group."[48] But elsewhere he makes clear that he believes some elements of our belief structure are so well established that we are unlikely to doubt them, and rightly so. Some elements of our work are well confirmed through other sources, or strongly linked to principles that we have little doubt are correct. Basic instruments such as thermometers and manometers, for example, are unlikely to be distrusted, as are the concepts that accompany them, such as temperature and pressure. Indeed, he insists that in testing the accuracy of a proposition, a physicist must make use of a whole group of theories that are accepted by him as "beyond dispute." Otherwise he would be paralyzed; it would be impossible for him to proceed. (One may suppose that basic principles of thermodynamics, such as conservation of mass and of energy, are in his mind.) Likewise if an experimental test fails, it does not tell us where the failure lies. It tells us only that somewhere in the system "there is at least one error."[49]

> In sum, the physicist can never subject an isolated hypothesis to experimental test, but only a whole group of hypotheses; when the experiment is in disagreement with his predictions, what he

learns is that at least one of the hypotheses constituting this group is unacceptable and ought to be modified; but the experiment does not designate which one should be changed.[50]

Duhem did not conclude that for this reason we should be radically skeptical. Rather he argued that we should adopt an attitude of reasonable humility toward intellectual commitments. Following Claude Bernard, he reminds us to be anti-dogmatic, to maintain an openness to the prospect that our theories may need revision, and to preserve an essential "freedom of mind."[51] Hypothesis, theories, and ideas in general are essential for stimulating our work, but we should not have "excessive faith" in them.[52] We should not be too pleased with our own accomplishments. As Americans at that time might have put it, we should not become "auto-intoxicated."[53]

In the face of an apparent refutation, how does a scientist decide which element(s) of the relevant nexus of theory, instruments, experimental setup, and auxiliary hypotheses should be revised? On this point, Duhem is not entirely satisfactory, invoking Pascal that there are "reasons which reason does not know." In the end, he concludes that these decisions ultimately are matters of judgment and "good sense."[54] Duhem uses history to underscore this point:

> We must really guard ourselves against believing forever warranted those hypotheses which have become universally adopted conventions, and whose certainty seems to break through experimental contradictions by throwing the latter back on more doubtful assumptions. The history of physics shows us that very often the human mind has been led to overthrow such principles completely, though they have been regarded by common consent for centuries as inviolable axioms, and to rebuild its physical theories on new hypotheses.[55]

Yet at the same time, he makes equally clear his conviction that history gives us grounds for confidence in the processes of scientific investigation, so long as we do not become dogmatic. He concludes with the following passage:

> The history of science alone can keep the [scientist] from the mad ambitions of dogmatism as well as the despair of . . . skepticism. By retracing for him the long series of errors and hesitations preceding the discovery of each principle, it puts him on guard against false evidence; by recalling to him the vicissitudes of the cosmological schools and by exhuming doctrines once triumphant from the oblivion in which they lie, it reminds him that the most attractive systems are only provisional representations, and not definitive explanations. And, on the other hand, by unrolling before him the continuous tradition through which the science of each epoch is nourished by the systems of past centuries . . . it creates and fortifies in him that conviction that physical theory is not merely an artificial system, suitable today and useless tomorrow, but that it is an increasingly more natural classification and an increasingly clearer reflection of realities which experimental method cannot contemplate directly.[56]

W.V.O. Quine and the Duhem-Quine Thesis

Duhem's views became known to American audiences primarily through the Harvard philosopher Willard Van Orman Quine, and in the process came to be viewed as more radical than they arguably were. Quine took the problem of refutation and reformulated it under the rubric of what has come to be known as "under-determination." If theories are tested not in isolation but in whole theoretical groups, then how do we know which piece of the group is in need of revision when something goes awry? Duhem's answer was: We rely on judgment. Quine's answer is: We *don't* know. Knowledge, he insists, is a web of belief. When we encounter a refutation, there is a universe of potential adjustments we can make, a universe of threads that can be tightened or loosened to sustain the fabric or reweave it. In Quine's words: "our statements about the external world face the tribunal of sense experience not individually but only as a corporate body."[57]

Duhem would have agreed with that, but he also believed that evidence could lead us to reexamine and adjust parts of that corporate body appropriately. This is one of his two key purposes of experimentation—to strengthen or weaken the support for particular elements in physical theory. If saving the phenomena required us to abandon something that is very strongly held—such as conservation of energy—we would be unlikely to do it. We would conclude that the experiment revealed a problem somewhere else or that there was a problem with our instrumentation. For Duhem, the various parts of the whole theoretical group are not created equal and not equally up for grabs. But Quine thinks that they are, concluding, famously: "any statement can be held true, come what may, if we make drastic enough adjustments elsewhere in the system."[58]

Quine's radical holism came to be known as the Duhem-Quine thesis and is taken by many scholars to weaken the grip of evidence on theory, because if theories are under-determined by experiment—and we have a world of choices in how to respond to experimental failure—then what is the basis for our belief?[59] It appears that some additional component is necessary to explain how scientists come to the conclusions that they do. This became the foundation of a great deal of what followed:

some scholars have argued that the concept of under-determination underpins the entire set of challenges to empiricist philosophy that developed in the second half of the twentieth century, including the work of Thomas Kuhn and emergence of the field of science studies.[60]

T. S. Kuhn and the Emergence of Science Studies

Thomas Kuhn's point of entry was to hoist the empiricists on their own petard: to assert that the empiricists have not been sufficiently empirical about science itself. His own work was grounded in the history of science through his early study of the Copernican Revolution—the topic of his first book—and his work at Harvard with James Conant developing a set of educational modules known the Harvard Case Histories in Experimental Science.[61] But Kuhn was also deeply engaged with arguments in philosophy of science and had read both Fleck and Quine, as well as works of the Vienna Circle.[62]

One of the central points of Kuhn's *Structure of Scientific Revolutions* was the same as Fleck's: scientists do not work alone but rather in communities that share not just theories about empirical reality—such as the theory of relativity or the theory of evolution by natural selection or the theory of plate tectonics—but also values and beliefs about how their science should operate. Together with models of exemplary scientific accomplishment ("exemplars"), these theories, values, and intellectual and methodological commitments collectively constitute the "paradigm" under which the community operates. This community aspect is paramount: in a 1979 forward to the first English translation of Fleck, Kuhn stressed that in the contemporary scientific world, a person working alone is more likely to be dismissed as a crank than accepted as a maverick.[63]

Most of the time, scientists do not question their paradigms, they work within them, solving problems and answering questions that the framework identifies as relevant. Kuhn called this "normal science" and asserted that its principal activities were a form of puzzle solving. Contra Popper, during normal science scientists do not attempt to refute the paradigm. In fact, they do not even question it—until a problem arises. This is where the engagement of science with reality becomes most evident: problems arise because some observation or experience of the world—some "technical puzzle"—does not fit expectation.[64] Kuhn calls these "anomalies." At first, scientists will attempt to account for the anomaly within the paradigm, perhaps making some modest adjustment in it. But if the anomaly becomes too great or too glaring, or the adjustments made to accommodate it generate new problems, this creates a crisis, which opens the intellectual space for reconsideration of the paradigm. Sometimes crises are resolved within the paradigm, but when they cannot be, a scientific revolution occurs: the governing paradigm is overthrown and replaced by a new one. It is like a political revolution, insofar as the new paradigm is in effect a new form of intellectual governance, with new rules and regulations. Kuhn thus argued that science advances neither by verification nor refutation, but by paradigm shifts.

Many scientists welcomed Kuhn's views insofar as they painted a picture of science that was recognizable to them, or at least more recognizable than the alternatives.[65] But what fired up the many readers who were not scientists was a claim that most scientists probably didn't understand and wouldn't have liked if they had (and what distinguishes Kuhn from Fleck): that successive paradigms are *incommensurable*. By this Kuhn meant, literally, that there was no metric by which a new paradigm could be compared to the one it proposed to replace. As Fleck had argued, the new paradigm—like the new thought-style—was not just a shift in thinking about a particular scientific question, it was also a shift in meanings, values, priorities, aspirations, and even the self-identity of the scientist. This opened still wider the question that Quine had posed: How do scientists decide which part of their belief structure needs to be revised in light of an anomaly? How do they decide whether a small adjustment is sufficient or a scientific revolution is in order? And if the new paradigm is incommensurable with the one it proposes to replace, on what basis do scientists make the choice to accept it?

Historians and philosophers have been debating these questions ever since. Philosophers were vexed by the incommensurability claim, insofar as it seemed to reduce paradigm choice to relativism and even irrationality.[66] Imre Lakatos, for example, opined that in Kuhn's theory, the scientific revolution is "a mystical conversion which is not and cannot be governed by rules of reason."[67]

Historians felt validated that Kuhn insisted on the detailed study of real science, but tended to find the incommensurability claim to be overblown, and noted that Kuhn had made a methodological error by sometimes comparing non-proximate scientific theories, such as Aristotelian physics and quantum mechanics. Yes, historians acknowledged, Aristotelian physics is inscrutable to a contemporary physicist, but there have been many intermediate steps between then and now; it does not work to try to understand the entire arc of the history of physics without tracing these intermediate steps. It would be like analyzing a relay race thinking that the baton had been thrown rather than passed.

My own view is that Kuhn was closer to the mark in his less famous earlier work *The Copernican Revolution*, in which he described a major scientific change as a bend in the road:

> From the bend, both sections of the road are visible. But viewed from a point before the bend, the road seems to run straight to the bend and disappear. . . . And viewed from a point in the next section, after the bend, the road appears to begin at the bend from which it runs straight on.[68]

Kuhn's work was itself a bend in the road of studies of science: away from method and toward practice; away from individuals and toward communities.[69] Scholars generally agree that the largest impact of Kuhn's work—besides adding the term *paradigm shift* to the general lexicon—was in helping to launch the field of science studies.

Away from Method

Philosophers from Comte to Popper attempted to identify the method of science that accounted for its success and therefore justified our acceptance of scientific claims as true—what is sometimes called "warranted true belief." Kuhn did not exactly say that there was no method, but he did say two things that displaced method from centrality. The first was the claim that under different paradigms, methods could change. The second was that most of the time, the methods of science amounted to not much more than puzzle-solving—working out details within the paradigm without questioning the larger structure—and that seemed pretty uninteresting. Moreover, whatever the methods were, they were done by groups of people working together, not individuals working alone.

This opened the door for an expanded sociology of science that not only examined the formal institutional structures of science, as previous sociologists had done, or the norms of scientific behavior, as the famous sociologist of science Robert Merton had studied, but addressed the *epistemological* question: What is the basis for scientific belief? If the intellectual action in science is in the paradigm shift, and if paradigms are incommensurable, then our traditional notions of scientific progress are clearly unsupportable. Perhaps science does not give us warranted true belief. Perhaps we should *not* trust science. If scientists can abandon one view and replace it with another incommensurable one, that does not inspire confidence in the idea that the processes of science necessarily provide us with a reliable view of the world. In any case, someone needs to explain the grounds on which scientists accept the claims they do.

Sociology of Scientific Knowledge and the Rise of Science Studies

Sociologists who took up the gauntlet thrown down by Kuhn called further attention to the social elements responsible for scientific conclusions, or what has come to be known as the *social construction* of scientific knowledge.[70] While they saw themselves as epistemological radicals, they were building on what had come before, particularly Quine's formulation of under-determination. They now asked: On what grounds do scientists decide what to believe and what to reject? How are these decisions articulated within the frameworks of scientific communities? To what degree, if any, should we respect the claims that emerge from this process?

The most influential of these early efforts came from the group of scholars we have come to know as the Edinburgh school, particularly Barry Barnes, David Bloor, and Steven Shapin. Barnes concentrated on "interests" as a driving force in theory choice. These "interests" could be professional, in the sense that the success of a favored theory would benefit the career of its promoter, or there could be an interest in a particular value set or a theory that was consistent with one's political, religious, or ethical views.[71] (In hindsight, interest theory seems oddly individualistic, but that is another matter.) Bloor insisted that the methods of science studies should be "symmetrical," meaning that "the same types of cause would explain, say, true and false beliefs."[72] Shapin attended particularly to the interrelationship between knowledge production and social order, arguing memorably, with historian Simon Schaffer, that "solutions to the problem of knowledge are solutions to the problem of social order."[73]

The arguments of the Edinburgh school were often taken to be ontologically anti-realist, and for that reason dismissed by many scientists as ridiculous.[74] To be sure, some scholars wrote in a manner that suggested a disregard for, if not outright disbelief in, the significance of empirical evidence in formulating scientific knowledge. It was easy to slip from the claim that empirical evidence does not by itself determine our conclusions to the suggestion that empirical evidence plays no role. But the argument was not so much anti-realist as it was *relativist*: if empirical evidence cannot determine decisively what we should believe and what we should reject, it does seem to suggest that our views are framed in relation to some set of standards and concerns that cannot be deduced from, nor reduced to, empirical evidence. And if social interests and conditions play a determinative role, then our knowledge must be at least in part relative to those interests and conditions. This was a very serious challenge. As Barnes explained in the 1970s, the approach of the Edinburgh school is

> sceptical since it suggests that no arguments will ever be available which could establish a particular epistemology or ontology as ultimately correct. It is relativistic because it suggests that belief systems cannot be objectively ranked in terms of their proximity to reality or their rationality.[75]

This was not the same as denying that our encounters with reality play a role in our convictions (much less to claim that there is no physical reality). Rather, it was to argue that the role of empirical evidence in shaping them was not nearly as determinative as most philosophers and scientists thought. Later commentators have generally allowed that the Edinburgh school was correct in stressing that evidence alone does not account for the conclusions to which scientists come.[76] The question, however, was whether Edinburgh theorists were suggesting that it played little or even no role. As Barnes allowed, "Occasionally, existing work leaves the feeling that reality has *nothing* to do with what is socially constructed or negotiated to count as natural knowledge, but we may safely assume that this impression is an accidental by-product of over-enthusiastic sociological analysis."[77]

This claim may be too generous; my own feeling is that some sociologists associated with or influenced by the Edinburgh school deliberately created this impression. When Karin Knorr-Cetina, for example, insisted in the 1980s that scientific knowledge was a "fabrication," when Harry Collins asserted that "the natural world in no ways constrains what is believed to be," and when Bruno Latour declared that science was "politics by other means," these terms and phrases were clearly chosen to unsettle what the historian John Zammitto has called the "ambient idolatry of science" that had prevailed under positivism.[78] More-over, by saying that "belief systems cannot be objectively ranked," Edinburgh scholars seemed to imply that objectivity did not play the role in science that scientists typically asserted, and perhaps played no role at all. These assertions were not acciden-tal; they were deliberate provocations.

But not all provocations are illegitimate, and the more impor-tant point, stressed recently by David Bloor, is that if we feel the need to contrast *relativism* with something, we should contrast it not with objectivity—which is the opposite of subjectivity—nor with truth, which is the opposite of falsehood—but with *absolutism*. The opposite of relative knowledge is absolute knowl-edge, and no serious scholar of the history or sociology of knowledge can sustain the claim that our knowledge is absolute. Nor can we sustain the claim that empirical evidence alone suf-fices to explain scientific conclusions. Far too much evidence refutes that hypothesis. Bloor has always been clear that he wants to be scientific in his study of science, and to be scientific about science means to take seriously the empirical evidence about the role of empirical evidence! And that empirical evidence reveals the limits of empiricism. Bloor's point has always been that when we look at science carefully and with an open mind, we see both empirical and social factors at play in stabilizing scientific knowl-edge, and we cannot assume a priori which ones are more important in any given case.[79]

A different critique of the notion of empirical method came from the philosopher Paul Feyerabend (1924–94). Born in Vienna, Feyerabend completed a PhD in philosophy on the topic of observation sentences and spent much of his life in con-versation with Karl Popper and Imre Lakatos, laying the groundwork for what might have been a career as a leading light of logical empiricism. But he later rejected not just logical empiricism, but any attempt to define or prescribe the method of science. In his most famous work, *Against Method* (published in 1975), he argued that there was no scientific method, nor should there be. Scientists have used a diversity of methods to good effect; any attempt to restrict this would hamper their creativity and impede the growth of scientific knowledge. Moreover, falsification as a rule is clearly falsified by the facts of history: few if any theories in the history of science ever explained all the available facts. Often scientists ignored facts that didn't fit or didn't seem significant, or set them aside to worry about at a later date.[80] (Popper might claim that those scientists were bad scientists, but if so then most scientists have been bad scientists, including some of our most celebrated.)

Like the science studies scholars quoted above, Feyerabend embraced a deliberately provocative style, and perhaps because he described his position as "theoretical anarchism" he is often quoted as having claimed that in science "anything goes." But that was not his claim. The actual quotation is this:

> It is clear then, that the idea of a fixed method, or a fixed theory of rationality, rests on too naïve a view of man and his social surroundings. To those who look at the rich material provided by history, and who are not intent on impoverishing it in order to please their lower instincts, their craving for intellectual security in the form of clarity, precision, "objectivity," [and] "truth," it will become clear that there is only one principle that can be defended under *all* circumstances and in all stages of human development. It is the principle: anything goes.[81]

Feyerabend was saying that if you *pressed* him to define the method of science, he would have to say that anything goes— which is to say that there is no unique method or principle of science. This was not an abdication of the responsibility to demarcate science from non-science, as Popper might have argued, but a recognition that methodological and intellectual diversity characterized the history of science, and this was a good thing: it made communities stronger, more creative, more open-minded, and nicer.[82] Absolutism—whether in science, politics, or anything else—was generally objectionable.[83] Like Popper (and Duhem and Comte), Feyerbend believed in progress; he just disagreed about whence it came. He summarized: "Theoretical anarchism is more humanitarian and more likely to encourage progress that its law-and-order alternatives . . . [and the] only principle that does not inhibit progress is: anything goes."[84] When we look seriously at what scientists do, we find that they are nothing if not creative, flexible, and adaptive.

Feyerabend was a philosopher, not a sociologist, and he accepted that science was progressive in a way that most of his sociological colleagues did not. But his work did support the sociological trend emerging strongly in the 1970s of focusing on the practices of scientists—in their labs, in the field, in clinical trials. If we cannot state a priori what the method of science is (or methods are), then the only way to find out is through observation.

The person who since then has done the most in that regard is unquestionably Bruno Latour, who turned the techniques of anthropology to science and in doing so drew particular attention to the practices that scientists employ to persuade their colleagues to accept any particular claim. Latour's great impact on the field was to establish ethnography as a key methodology in science studies, and to insist on the importance of privileging what scientists do over what they say.[85] While the work that has followed in his wake defies easy summation, one thing is clear: it confirms earlier arguments about scientific methodological diversity. After the work of the Edinburgh school, of Feyerabend, of Latour and his colleagues, and of the diverse historians who have documented the ways scientific methods have changed over time, it is no longer plausible to hold to the view that there is any singular scientific method.[86]

This is not an entirely negative finding, but it does commit us to the conclusion that the dream of positive knowledge has truly ended.[87] There is no identifiable (singular) scientific method. And if there is no singular scientific method, then there is no way to insist on ex ante trust by virtue of its use. Moreover, despite the claims of prominent scientists to the contrary, the contributions of science cannot be viewed as permanent.[88] The empirical evidence gleaned from the history of science shows that scientific truths are perishable. How can we tell then if scientific work is good work or not? On what basis should we trust or distrust science?

Getting Unstuck: Social Epistemology

Despite the challenges of science studies, there have still been many attempts to salvage scientific rationality. In my view, the most successful of these have come from a direction that most scientists would have least suspected: feminism.

Since the 1960s, feminists have asked: How could science claim to be objective when it largely excluded half the population from the ranks of its practitioners? How could science claim to be producing disinterested knowledge when so many of its theories embedded obvious social prejudices, not just about gender but also about race, class, and ethnicity? These questions were not necessarily hostile. Many of them were raised by female scientists who were interested in the natural or social world and believed in the power and value of scientific inquiry to explain it.

Sociologists of scientific knowledge stressed that science is a social activity, and this has been taken by many (for both better and worse) as undermining its claims to objectivity. The "social," particularly to many scientists but also many philosophers, was synonymous with the personal, the subjective, the irrational, the arbitrary, and even the coerced. If the conclusions of scientists—who for the most part were European or North American men—were social constructions, then they had no more or less purchase on truth that the conclusions of other social groups. At least, a good deal of work in science studies seemed to imply that.

But feminist philosophers of science, most notably Sandra Harding and Helen Longino, turned that argument on its head, suggesting that objectivity could be reenvisaged as a social *accomplishment*, something that is collectively achieved.[89] Harding mobilized the concept of *standpoint epistemology*—the idea that how we view matters depends to a great extent on our social position (or, colloquially, that where we stand depends on where we sit)—to argue that greater diversity could make science stronger. Our personal experiences—of wealth or poverty, privilege or disadvantage, maleness or femaleness, heteronormativity or queerness, disability or able-bodiedness—cannot but influence our perspectives on and interpretations of the world. Therefore, *ceteris paribus*, a more diverse group will bring to bear more perspectives on an issue than a less diverse one.[90]

In her groundbreaking 1986 book, *The Science Question in Feminism*, Harding argued that the objectivity practiced by most scientific communities was weak, because of the characteristic homogeneity of those communities. The perspectives of women, people of color, the working classes, and many others were lacking, and the consequences were plain to see when one considered the obvious sexism, racism, and class bias of many past scientific theories. But there could be less obvious forms of bias at work as well. She argued for what she labeled *strong objectivity*: an approach that acknowledged that an individual's beliefs, values, and life experiences necessarily affect their work—scientific or otherwise—so the best way to develop objective knowledge is to increase the diversity of knowledge-seeking communities. Objectivity was not a 0/1 proposition: communities could be more or less objective and greater objectivity in scientific research achieved—or at least made more likely—by greater heterogeneity in the scientific community.[91]

Like Feyerabend, Harding tended toward the deliberately provocative—as when she compared Newton's *Principia Mathematica* to a rape manual—and this made her an easy target of right-wing critics.[92] It also made her the target of scientific critics, such as Paul Gross and Norman Levitt, who failed to understand that the central point of her critique was that science could be made stronger through inclusion. This point was made a bit more diplomatically—albeit equally forcefully from an intellectual standpoint—by the feminist philosopher Helen Longino.

Longino transformed a common scientific assumption—that science *is* self-correcting—into a pressing intellectual question— *How* is it that science is self-correcting? After all, the claim that science corrects itself might be viewed as highly implausible—a sort of epistemic magic trick. Longino's suggested that it is not so much that *science* corrects *itself*, but that *scientists* correct *each other* through the social processes that constitute "transformative interrogation." It is through the give and take of ideas—the challenging, the questioning, the adjusting and amending—that scientists integrate their colleagues' work, offer up criticisms, and contribute to the growth of warranted knowledge. She wrote:

> The objectivity of individuals in this scheme consists in their participation in the collective give-and-take of critical discussion and not in some special relation (of detachment, hardheadedness) they may bear to their observations. Thus understood, objectivity is dependent upon the depth and scope of the transformative interrogation that occurs in any given scientific community.[93]

Longino urged us to accept (rather than lament) the fact that individual scientists invariably bring biases, values, and background assumptions into their work. The scientist entering the laboratory cannot hang up her personal values, preferences, assumptions, and motivations like an overcoat, as Claude Bernard once supposed.[94] What can happen, however, is that in a diverse community subjective elements can (and most likely will) be challenged by others, and to the extent that they may be inappropriately informing evidential reasoning and theory choice, that can be challenged, too.[95]

Longino's account of transformative interrogation solves the problem of how science, as a whole, can be objective even when individual scientists are not:

> If scientific inquiry is to provide knowledge, rather than a random collection of opinions, there must be some way of minimizing the influence of subjective preferences and controlling the role of background assumptions. The social account of objectivity solves this problem. The role of background assumptions in evidential reasoning is grounds for unbridled relativism only in the context of an individualistic concept of scientific method and scientific knowledge. . . . Values are not incompatible with objectivity, but objectivity [emerges] as a function of community practices rather than as an attitude of individual researchers.[96]

This perspective reinforces Harding's position that objectivity is not a matter of either/or, but of degree. The greater the diversity and openness of a community and the stronger its protocols for supporting free and open debate, the greater the degree of objectivity it may be able to achieve as individual biases and background assumptions are "outed," as it were, by the community. Put another way: objectivity is likely to be maximized when there are recognized and robust avenues for criticism, such as peer review, when the community is open, non-defensive, and responsive to criticism, and when the community is sufficiently diverse that a broad range of views can be developed, heard, and appropriately considered. On this view, it is not surprising that when scientists were almost exclusively white men, they developed theories about women and African Americans that were at best incomplete and at times pernicious—theories that have now been rejected. Nor is it surprising that many of the logical and empirical flaws of these earlier theories were pointed out by women and people of color.[97] (This point is addressed further in chapter 2.)

The key point here is that often "assumptions are not perceived as such."[98] They are so embedded as to go unrecognized *as* assumptions, and this is most likely to occur in homogeneous communities. Longino continues:

When, for instance, background assumptions are shared by all members of a community, they acquire an invisibility that renders them unavailable for criticism. They do not become visible until individuals who do not share the community's assumptions can provide alternative explanations of the phenomena without those assumptions, as, for example, Einstein could provide an alternative explanation of the Michelson-Morley interferometer experiment [because he did not share the assumption of the variable speed of light]. . . . From all this it follows again that the greater the number of different points of view included in a given community, the more likely it is that its scientific practice will be objective . . . [and] it will result in descriptions and explanations of natural processes that are more reliable . . . than would otherwise be the case.[99]

Transformative interrogation can empower us to decide whether those background assumptions are, in a given context, appropriate and helpful or inappropriate and unhelpful. This is most likely to occur in a diverse community for the simple reason that diverse communities will have diverse background assumptions. Diversity does not heal all epistemic ills, but *ceteris paribus* a diverse community that embraces criticism is more likely to detect and correct error than a homogeneous and self-satisfied one.[100]

Feminist epistemology soundly refutes the claim that the social character of science makes it subjective. On the contrary, we can now see that scientists who were offended by the social turn in science studies—as well as science studies scholars who thought they could debunk science by exposing its social character—got it wrong. The feminist account of the social character of science can make a stronger case for the objectivity of scientific knowledge than previous accounts by identifying both sources of bias and remedies to it. And consider this: in their dyspeptic polemic of the 1990s, *Higher Superstitions: The Academic Left and Its Quarrels with Science,* scientists Paul Gross and Norman Levitt accused feminists of being anti-science. But neither Harding nor Longino were anti-science.[101] Both were discussing ways to strengthen and improve it. Gross and Levitt could have used feminist philosophy of science in their defense of science had they not been so busy taking offense.

In Diversity There Is
Epistemic Strength

Feminist philosophy of science salvages science from the claim that its social character makes it subjective, but it does leave us with a view of science that makes some people uncomfortable: that science is fundamentally *consensual*. Longino summarizes: "To say that a theory or hypothesis was accepted on the basis of objective methods does not entitle us to say it is true but rather that it reflects the critically achieved consensus of the scientific community. [And] it's not clear we should hope for anything better."[102] I agree. But where does that leave us?

To recapitulate: There is now broad agreement among historians, philosophers, sociologists, and anthropologists of science that there is no (singular) scientific method, and that scientific practice consists of communities of people, making decisions for reasons that are both empirical and social, using diverse methods. But this leaves us with the question: If scientists are just people doing work, like plumbers or nurses or electricians, and if our scientific theories are fallible and subject to change, then what is the basis for trust in science?

I suggest that our answer should be two-fold: 1) its sustained engagement with the world and 2) its social character.

The first point is crucial but easily overlooked: Natural scientists study the natural world. Social scientists study the social world. That is what they do. Consider a related question: Why trust a plumber? Or an electrician? Or a dentist or a nurse? One answer is that we trust a plumber to do our plumbing because she is trained and licensed to do plumbing. We would *not* trust a plumber to do our nursing, nor a nurse to do our plumbing. Of course, plumbers can make mistakes, and so we get recommendations from friends to ensure that any particular plumber has a good track record. A plumber with a bad track record may find herself out of business. But it is in the nature of expertise that we trust experts to do jobs for which they are trained and we are not. Without this trust in experts, society would come to a standstill. Scientists are our designated experts for studying the world.[103] Therefore, to the extent that we should trust anyone to tell us about the world, we should trust scientists.

This is not the same as faith: We do (or should) check the references of our plumbers and we should do the same for our scientists. If a scientist has a track record of error, underestimation, or exaggeration, this might be grounds for viewing his or her claims skeptically (or at least judging their results with this information in mind.) If a scientist is receiving financial support—directly or indirectly—from an interested party, this may be grounds for applying a higher level of scrutiny than we might otherwise demand. (For example, an editor might send the paper for additional review, or a reviewer might pay extra attention to study design, where subconscious bias may slip in.)[104]

No doubt individual scientists, like individual plumbers, may be stupid, venal, corrupt, or incompetent. But consider this: the *profession* of plumbing exists because in general plumbers do a job we need them to do, and in general they do it successfully. When we evaluate the track record of science, we find a substantial record of success—in explanation, in prediction, in providing the basis for successful action and innovation. We have a world of medicines, technologies, and conceptual understandings derived from science that have enabled people to do things they have wanted to do. (As already noted, that success does not prove that the theories involved are necessarily true, but it does suggest that scientists are doing something right.) This might be the one point on which the diverse scholars I have discussed agree: philosophers, historians, sociologists, and anthropologists have all been interested in science because of its *success*—both culturally and epistemologically. The question of this lecture is of interest at least in part because the success of science as a source of stable epistemic authority has been called into question, and its future success as a cultural enterprise appears to be at least somewhat in doubt.

This consideration—that scientists are in our society the experts who study the world—is a reminder to scientists of the importance of foregrounding the empirical character of their work—their engagement with nature and society and the empirical basis it provides for their conclusions. As I have stressed elsewhere, scientists need to explain not just what they know, but how they know it.[105] Expertise as a concept also carries with it the embedded idea of specialization, and therefore the limits to expertise, reminding us why it is important for scientists to exercise restraint with respect to subjects on which they lack expertise.

However, reliance on empirical evidence alone is insufficient for understanding the basis of scientific conclusions and therefore insufficient for establishing trust in science. We must also take to heart—and explain—the social character of science and the role it plays in vetting claims. Here it is worth reiterating my point that scientists who were offended by the "social" turn in science studies got it wrong: much of what we identify as "science" are social practices and procedures of adjudication designed to ensure—or at least to attempt to increase the odds—that the process of review and correction are sufficiently robust as to lead to empirically reliable results.[106] Again, Longino: "Socializing cognition is not a corruption or displacement of the rational but a vehicle of its performance."[107]

Peer review is one example of such a practice: it is through peer review that scientific claims are subjected to critical interrogation. (This is why, in my own work, I have stressed the importance of evaluating scientific consensus through analysis of the peer-reviewed literature and not the popular press or social media, and why these chapters were subject to peer review.) This includes not only the formal review that papers go through when submitted to academic journals, but also the informal processes of judgment and evaluation that research findings undergo when scientists discuss their preliminary results in conferences and workshop and solicit comments from colleagues prior to submitting them for publication, as well as the continued process of evaluation that published claims endure as fellow scientists attempt to use and build on those claims.[108]

Tenure is another example: we evaluate scholars' work in order to judge whether they are worthy of joining the community of scholars in their fields, in effect to be certified as experts. Tenure is effectively the academic version of licensing. The crucial element of these practices is their social and institutional character, which work to ensure that the judgments and opinions of no one person dominate and therefore that the value preferences and biases of no one person are controlling. Of course, within any community there will be dominant groups and individuals, but the social processes of collective interrogation offer a means for the less dominant to be heard so that, to the maximum degree possible, the conclusions arrived at are non-partisan and non-idiosyncratic.[109] The social character of science forms the basis of its approach to objectivity and therefore the grounds on which we may trust it.

In recent years, this insight has been implicitly incorporated into scientific practices, particularly in just those domains where scientific claims are likely to be viewed as controversial.

The US National Academy of Sciences works to ensure that the panelists who perform its reviews are diverse and represent a range of viewpoints. Scholars have called this approach the "balancing of bias."[110] The Intergovernmental Panel on Climate Change—now one of the world's largest aggregations of scientists—makes a particular point of seeking geographical, national, racial, and gender diversity in its chapter-writing teams. While the motivations for inclusivity may be in part political, the widespread character of practices of inclusion suggest that many scientific communities now recognize that diversity serves epistemic goals.

Caveats

My arguments require a few caveats. Most important is that there is no guarantee that the ideal of objectivity through diversity and critical interrogation will always be achieved, and therefore no guarantee that scientists are correct in any given case. The argument is rather that, given the existence of these procedures and assuming they are followed, there is a mechanism by which errors, biases, and incompleteness can be identified and corrected. In a sense, the argument is probabilistic: that if scientists follow these practices and procedures, they increase the odds that their science does not go awry. Moreover, outsiders may judge scientific claims in part by considering how diverse and open to critique the community involved is. If there is evidence that a community is not open, or is dominated by a small clique or even a few aggressive individuals—or if we have evidence (and not just allegations) that some voices are being suppressed—this may be grounds for warranted skepticism. In this respect, each case must be evaluated on its own merits.

An interesting recent case is the "extended evolutionary synthesis" (EES) concept, which challenges the primacy of genetic control in inheritance and calls increased attention to developmental plasticity, environmental modification by organisms (including niche construction), epigenetics, and social learning.[111] Some advocates of EES have been disturbed by resistance they have encountered among "traditionalists" in the evolutionary biology community, who argue that the existing evolutionary synthesis is adequate and no extension is needed.[112] The ensuing arguments have sometimes become hostile and personal.[113] To a historian familiar with past major debates in science, it is not surprising that there is resistance to new ideas that threaten the stability of past scientific achievements or the social position of their adherents, nor that this resistance at times gets heated.[114] When people's life work is being questioned, they may get testy. No one likes to be told that they are wrong. The important question here is whether the advocates of EES have been able to publish their views in respected journals and to obtain funding for their research. The answer is yes. Despite the flaring of tempers, the evolutionary biology community as a whole has proved open to the introduction of new ideas and the critical interrogation of old ones.

A second caveat is that my argument is by no means a call for blind or blanket trust, much less a slavish adherence to scientists' recommendations on non-scientific matters. It is a call for *informed* trust in the consensual conclusions of scientific communities, but not necessarily in the views or opinions of individual scientists, particularly not when they stray outside their domains of expertise. Indeed, the track record of scientists outside their specialties is not particularly impressive. One need only think of physicist-mathematician John von Neumann claiming in the 1950s that within a few decades nuclear energy would be as free as the "unmetered air," or physicist William Shockley's insistence that African Americans were genetically inferior to whites and should be paid to undergo "voluntary" sterilization.[115] Werner von Braun thought that by the year 2000, the first child would have been born on the moon.[116] Physical scientists, particularly in the United States, have tended to be technofideists, exaggerating the rate at which new technologies would be developed or the degree to which they would improve our lives. Both physical and life scientists have an unhappy record of insensitivity to social and ethical concerns, as witnessed by the widespread support among biologists in the early twentieth century for eugenics programs that in hindsight appear both scientifically erroneous and morally noxious (see chapter 2). Outside their domains of expertise, scientists may be no more well informed than ordinary people. Indeed, they may be less so as their intense training in one area can lead them to be undereducated in others.[117]

The claim that scientists have expertise in particular domains is not, moreover, to insist that this expertise is exclusive. Many lay people—farmers and fishermen, midwives and patients—have expertise in their particular domains.[118] Patients may have considerable understanding of the progression of their disease or the side effects of pharmaceuticals; midwives may be able to recognize problems in pregnancies as well or better than some obstetricians. There was extensive scientific knowledge in India before the arrival of the British, particularly about matters that the British would label "natural historical" (but locals might not have labeled this way).[119]

We have a considerable literature on indigenous expertise: the knowledge that both lay people and experts may have about plants, animals, geography, climate, or other aspects of their natural environments and communities. In recent decades we have come to understand more fully the empirical knowledge systems that have developed outside of what we conventionally call "Western science"—what anthropologist Susantha Goonatilake has called "civilizational knowledge." These systems may involve highly developed expertise, and may be quite effective in their realms.[120] For example, Traditional Chinese Medicine (TCM), acupuncture, and Ayurvedic medicine can be efficacious in treating certain diseases and conditions for which Western medicine has little to offer.[121] Civilizational knowledge traditions have authority in their regions of origin by virtue of track records of success, and in some cases (e.g., acupuncture) have demonstrated efficacy beyond those regions as well. Moreover, the study of civilizational knowledge has highlighted the values embedded in Western science that often go unrecognized or are even denied by its practitioners.[122]

There are also lay knowledge traditions based on sustained empirical and analytical engagement with the world. Hunter-gatherer societies, for example, typically have detailed empirical knowledge of plant distributions and animal migrations; anthropologist Colin Scott, for example, has demonstrated that Cree hunting traditions are highly empirical, and argues that they are therefore rightly viewed as scientific.[123] Where lay knowledge overlaps with scientific knowledge, one should not assume that the latter is necessarily superior to the former.[124] We know, for example, that Polynesian navigators were far more effective in plying the Pacific than their European counterparts until at least the time of Cook in the late eighteenth century.[125]

There is an important distinction to be made here: respecting indigenous, lay, and "Eastern" knowledge that has demonstrated empirical adequacy or clinical efficacy is a very different thing from accepting popular claims that are ignorant, erroneous, or represent motivated disinformation. The claims of an actress that vaccines cause autism or an oil executive that recently observed climate change has been caused by sunspots do not come out of established knowledge traditions; the individuals promoting them do not have a credible claim to expertise. An actress is not an immunologist; a petroleum industry CEO is not a climate scientist. And in these particular cases, we have abundant empirical evidence that their claims are untrue. The claim that climate change is caused by sunspots has had its day in scientific court: it has been vetted by evidence and shown to be incorrect.[126] Autism is no more common among children who have been vaccinated than those who have not.[127] Respecting alternative knowledge traditions does not mean that we suspend judgment, either about those traditions or our own.

It is also important to distinguish between the scientific and the normative questions that get mooted in contemporary society. To be sure, the interrelations between the various sciences and the politics, economics, and morality that surround and embed them are often complex, intercalated, and not easily disentangled; some scholars have argued that they cannot be disentangled.[128] I believe that, however imperfectly, we can distinguish between the scientific and normative aspects of many questions—and that we continue to need to. Whether manmade climate change is underway is a different sort of question from what we should do about it; I may have reasons for declining vaccination that have nothing to do with its alleged relation to autism.[129] These distinctions matter, because if I understand that some of my fellow citizens reject vaccinations on religious grounds, I may respect that opinion without succumbing to the fallacy that vaccines cause autism; depending on my own religious views, I might join them or I might not. Similarly, I can respect the fact that many people have had adverse reactions from pharmaceuticals and know that iatrogenic illness is a real thing, without accepting the allegation that it is the drug AZT, rather than a virus, that causes HIV-AIDs.[130] Pope Francis rejects genetically modified organisms as an inappropriate interference with God's domain; if I were Catholic I might choose to follow his views irrespective of the scientific evidence as to whether those products are safe to eat.[131] Distinctions between the scientific and the social matter, because they rightly affect our choices, and because us they help to distinguish between arguments that may be persuasive to our audiences and arguments that are doomed to fail because they don't address their underlying concerns.

Comte argued long ago that the basis for the success of science was experience and observation. We now know that that is only part of the story, albeit an important part. Nevertheless, we can use this argument to remember that the basis for *our* trust in science is, in fact, experience and observation—not of empirical reality, *but of science itself*. It is what Comte argued long ago: that just as we can only understand the natural world by observing it, so we can only understand the social world by observing it. When we observe scientists, we find that they have developed a variety of practices for vetting knowledge—for identifying problems in their theories and experiments and attempting to correct them. While these practices are fallible, we have substantial empirical evidence that they do detect error and insufficiency. They stimulate scientists to reconsider their views and, as warranted by evidence, to change them. This is what constitutes progress in science.

Coda: Why Not the
Petroleum Industry?

We can now answer the question raised at the outset of this chapter of ex ante trust. Why should the conclusions of climate scientists about climate change be viewed ex ante as more authoritative than those originating from the petroleum industry? Or arguments about cancer and heart disease from the tobacco industry? Or about diabetes or obesity from Coca-Cola?[132]

The answer is simple: conflict of interest. The petroleum industry exists to explore for, find, develop, and sell petroleum resources, and by doing so to make a profit and return value to shareholders. It relies heavily on science and engineering to do this, and company scientists and executives have considerable expertise in the domains of sedimentary geology, geophysics, and petroleum and chemical engineering, as well as sales and marketing. But recent scientific findings about the reality and severity of anthropogenic climate change—and the role of greenhouse gases derived from fossil fuel combustion in driving it—threaten not only the industry's profitability, but even its existence. The fossil fuel industry as we know it is fighting for its survival. Rather than accept the necessity of change, certain elements in the industry have misrepresented the scientific evidence that demonstrates that necessity.[133] Exxon Mobil may be a reliable source of information on oil and gas extraction, but it is unlikely to be a reliable source of information on climate change, because the former is its business and the latter threatens it.[134]

We may say the same about the tobacco industry. For decades, the tobacco industry refused to accept the scientific evidence that tobacco products caused cancer, heart disease, bronchitis, emphysema, and a host of serious conditions and fatal diseases, including sudden infant death syndrome. It worked to challenge, discredit, and suppress known information, and it paid scientists to engage in research that was in other respects legitimate, but whose purpose (from the industry standpoint) was to distract attention from the adverse effects of tobacco use. The chemical industry has done much the same with respect to pesticides and endocrine disrupting chemicals; in recent years we have seen some of the same strategies and tactics taken up by elements of the processed food industry.[135] The tobacco, processed food, and chemical industries face an essential conflict of interest when discussing scientific results that bear on the safety, efficacy, or healthfulness of their products. They are not engaged in good faith in the open, critical, and communal vetting of evidence that is crucial for the determination of the reliability of scientific claims. This is why, ex ante, we have reason to distrust them.

This is not to say that an individual scientist or team of scientists is necessarily discredited simply because they work in or for a potentially conflicted industry or have received funding from it. Scientists within an industry may participate in the scientific enterprise by doing research and submitting it for publication in peer-reviewed journals, and there are many fine examples of this, particularly in the early twentieth century when many corporations ran large industrial research laboratories. (Full disclosure: my own PhD work was partly funded by the mining company for whom I worked before going to graduate school, and this was disclosed in my relevant publications.)

When industry-funded scientists attend conferences and publish in peer-reviewed journals, they are acting as parts of scientific communities, participating in the norms of those communities and subjecting themselves and their work to critical scrutiny. As long as they do—so long as the norms of critical interrogation are operating and conflicts of interest are forth-rightly disclosed and where necessary addressed—these scientists may well make fine contributions.[136]

But it is scarcely a secret that the goals of profit-making can collide with the goals of critical scrutiny of knowledge claims. We know from history that industrial research can be of high quality, but we also know that it exists—and is subject to external scrutiny—at the discretion of the industrial sponsor. Excellent research has emerged from the precincts of American business and industry, but so has disinformation, misrepresentation, and distraction. Science done within industries has won Nobel prizes; it has also been subject to suppression and distortion. Moreover, as Robert Proctor, Allan Brandt, David Rosner, Gerald Markowitz, Miriam Nestle, Erik Conway, and I have documented, a substantial amount of industry research has been *designed* to be a distraction.[137] Empirical reality tells us that we are right to be suspicious when the petroleum industry makes claims about climate science or the soda industry offers up nutritional claims, just as we should have been suspicious when the tobacco industry told us that Luckies were good for us and Camels would aid our digestion.[138]

The checkered history of scientific research in American industry that was designed to distract, confuse, and/or misinform also helps us to address one of the more nefarious strategies of industry doubt-mongering: the claim that they are instantiating the spirit of scientific inquiry when they pose skeptical questions and that it is *scientists* who are being dogmatic. This is an intellectually noxious move, because it takes the strength of science and turns it into a weakness, and falsely imputes scientific motives to activities that are intended to undermine science. Moreover, when scientists are unfairly attacked, they may become defensive and therefore less open to warranted critique than they should be. In this regard doubt-mongering is doubly damaging: it undermines public trust in science and it has the potential to undermine science itself.

The processes of critical interrogation rely on an assumption of good faith: that participants are interested in learning and have a shared interest in truth. It assumes that the participants do not have an intellectually compromising conflict of interest. When these assumptions are violated—when people use skepticism to undermine and discredit science rather than to revise and strengthen it, and to confuse audiences rather than to inform them—the entire process is disabled.[139] It can lead scientists to want to shut down criticism rather than embrace it. After all, it is challenging to maintain a spirit of openness in the face of dishonesty. The critics of science do not strengthen it, as they sometimes claim; they damage it.

For this and many other reasons, there is no guarantee that the methods of scientific scrutiny will operate as intended. In the next chapter, I examine historical examples where, in hindsight, we may say that scientists went astray. We will see what we may be able to learn from those examples as to when we are justified in not trusting science. But for the purposes of the present argument, the key point is this: We have an overall basis for trust in the processes of scientific investigation, based on the social character of scientific inquiry and the collective critical evaluation of knowledge claims. And this is why, ex ante, we are justified in accepting the results of scientific analysis by scientists as likely to be warranted.

Chapter 2

SCIENCE AWRY

If you google "How old is the Earth?" the first answer you will be offered is 4.543 billion years old. This is the accepted scientific value based on radiometric dating of asteroids and lunar materials. You will find it affirmed if you visit the web page of NASA, the US Geological Survey, or the Encyclopedia Britannica. It has been more or less in place for half a century and most educated Americans accept it as factual. It is what any mainstream earth science professor or teacher would teach, and what you will find in any college earth science textbook. However, if you scroll down on your computer you will also find:

> How old is the earth?—creation.com
> *creation.com/how-**old-is-the-earth***
> Creation Ministries International

The answer offered by Creation Ministries International, based upon biblical exegesis, is about six thousand years. If we were to judge a knowledge claim by the antiquity of its provenance, we would have to judge this claim to be more stable than the accepted scientific one, because it has been around since the mid-seventeenth century. Similarly, if we were to define authority as the ability to drive out competing claims, then the authority of the scientific value is clearly by no means total. This is not merely the case for the age of the Earth. If we look for answers about climate change, the safety of vaccinations, whether plate tectonics is an accurate and adequate theory of global tectonics, and if drinking water fluoridation prevents cavities, we will find many claims competing for our attention.

Some of these claims are simply unscientific—which is to say not based on vetted evidence—while others have been shown by evidence to be false. Yet they persist. Indeed, the fragile status of facts—both scientific and social—is now so widely acknowledged that the Oxford English Dictionaries declared the 2016 word of the year to be "post-truth."[1] Comedian Stephen Colbert complained that this was a rip-off of his earlier neologism "truthiness."[2]

The inclination of some religious believers to distrust scientific findings is neither new nor unstudied. Scholars have amply described and attempted to account for religiously motivated dissent from scientific theories of evolution from Darwin to Dawkins. But rejection of scientific claims is not restricted to matters of theological concern; people reject scientific conclusions for a host of reasons. Clearly, the establishment of scientific claims qua science does not entail the acceptance of those claims by people outside the scientific community. On the contrary, a "post-truth" world is one in which the fundamental assumptions of scientific inquiry—including its capacity to yield objective, trustworthy knowledge—have been called into question.

Some scholars, most notably Bruno Latour and Sheila Jasanoff, have argued that scientific knowledge is co-produced by scientists and society, in which case truthiness might be viewed as a normal state of affairs.[3] A co-produced claim, in their view, is one on which both scientists and society have converged, and it is this convergence—rather than empirical reality or even empirical support—that grants stability to the claim. Until this scientific and social convergence occurs, disputation is inevitable, and not just about values but also about facts.

As an empirical matter, this is clearly so. But the concept of co-production begs the question of what it means for a claim to be scientific and whether factual claims should fairly be understood as distinct from other types of claims. It also begs the question of whether we are justified in rejecting (or at least suspending judgement on) a claim that scientists consider settled when other members of society have demurred. The theory of co-production begs the question of whether scientific claims made by scientific experts merit trust.[4]

Latour has argued that scientific claims are *performances* about the natural world, and that scientists have been successful at "performing the world we live in."[5] By this he (presumably) means that scientists have achieved substantial social authority and are broadly accepted as our leading societal experts on "matters of fact."[6] (They perform and we applaud.) He also suggests (presumably ruefully) that natural scientists are "better equipped at performing the world we live in than [social scientists] have been at deconstructing it."[7] But he may be overestimating the success (performative or otherwise) of the natural sciences, given the large numbers of Americans who doubt many important claims of contemporary science (I restrict myself to the United States here, but similar claims could be made about other countries, such as the HIV-AIDS link in parts of Africa).

If we define success in terms of cultural authority, the success of science is clearly not only incomplete but at the moment looking rather shaky. Large numbers of our fellow citizens—including the current president and vice president of the United States—doubt and in some cases actively challenge scientific conclusions about vaccines, evolution, climate change, and even the harms of tobacco. These challenges cannot be dismissed as "scientific illiteracy." Studies show that in the United States, among Democrats and independent voters, higher levels of education are correlated with higher levels of acceptance of scientific claims, but among Republicans the opposite is true: The more educated Republicans are, the more likely they are to doubt or reject scientific claims about anthropogenic climate change. This indicates not a lack of knowledge but the effects of ideological motivation, interpreted self-interest, and the power of competing beliefs.[8]

And, as we saw in chapter 1, there is a deeper problem, one that transcends our particular political moment and varying cultural conditions. Even if we accept contemporary scientific claims as true or likely to be true, history demonstrates that the process of transformative interrogation will sometimes lead to the overturning of well-established claims. William James argued more than a century ago that experience has a "way of boiling over, and making us correct our present formulas."[9] He astutely pointed out that what we label as "'absolutely' true, meaning what no further experience will ever alter, is that ideal vanishing point toward which we imagine that all our temporary truths will someday converge . . . Meanwhile we live today by what truth we can get today, and be ready tomorrow to call it falsehood."[10] This was also Karl Popper's point where he argued for the provisional character of all scientific knowledge.

The overturning of claims is not arbitrary; it is related to experience and observation. But why we should accept any contemporary claim if we know that it may in the future be overturned? One might point out that incomplete and even inaccurate knowledge may still be useful and reliable for certain purposes: the Ptolemaic system of astronomy was used to make accurate predictions of eclipses, and airplanes were flying before aeronautical engineers had an accurate theory of lift.[11] That scientific knowledge may be partial or incomplete—or that old theories get replaced by new ones—is not ipso facto a refutation of science in general. On the contrary, it may be read as proof of the progress of science, particularly when in hindsight we can look back on the older theories and understand how and why they worked. (Newtonian mechanics still works when the objects under consideration are not moving very quickly.) But if our knowledge is overturned wholesale—if it is deemed in hindsight to have been wholly incorrect—that calls into question whether we can trust current scientific knowledge when we need to make decisions.[12]

Climate skeptics sometimes raise this point. In public lectures on climate science, I have been asked: "Scientists are always getting it wrong, so why should we believe them about climate change?" The "it" that scientists are allegedly getting wrong is rarely specified, and when I ask my interlocutor what he has in mind, usually there is no specific answer. When there is, most often it is the changing and seemingly contradictory recommendations of nutritionists. There are many reasons why nutritional information in recent years has been a moving target, and why nutrition seems to be a dismal science. These include the role of the mass media in publicizing novel but unconfirmed findings; the misuse of statistics by ill-trained scientists; the problems of small sample size and the difficulty of undertaking a controlled study of people's eating habits (see Krosnick, this volume); and the influence of the food industry in funding distracting research on the relative harms of sugar and fat.[13] (Elsewhere I have written on the potential adverse effects of industry funding of science when the desired outcomes are clear and biasing.[14]) But even if nutritional science is atypical, or even if it is typical but the sources of confusion in it can be identified and addressed, the skeptical challenge is epistemologically legitimate.

If scientists sometimes get things wrong—and of course they do—then how do we know they are not wrong now? Can we trust the current state of knowledge?

In this chapter, I set aside the issues of corruption, media misrepresentation, and inadequate statistical training to look at a problem that I think is more vexing, and certainly more challenging epistemically. It is the problem of science gone awry of its own accord. There are numerous examples in the history of science of scientists coming to conclusions that were later overturned, and many of those episodes have to do neither with religious commitments, nor overt political pressures, nor commercial corruption.[15] This has been the central question guiding much of my research career: How are we to evaluate the truth claims of science when we know that these claims may in the future be overturned?

Elsewhere I have called this problem the instability of scientific truth.[16] In the 1980s, philosopher Larry Laudan called it the pessimistic meta-induction of the history of science.[17] He observed (as have many others) that the history of science offers many examples of scientific "truths" that were later viewed as misconceptions. Conversely, ideas rejected in the past have sometimes been rescued from their epistemological dustbins, brushed off, polished up, and accepted into the halls of respectable science. The retrieval of continental drift theory and its incorporation into plate tectonics—the topic of my first book— is a case in point.[18] As I wrote in 1999 when discussing that retrieval: "History is littered with the discarded beliefs of yesterday and the present is populated by epistemic resurrections." Given the perishability of past scientific knowledge, how are we to evaluate the aspirations of contemporary scientific claims to legitimacy and even permanence?[19] For even if some truths of science prove to be permanent, we have no way of knowing which ones those will be. We simply do not know which of our current truths will stay and which will go.[20] How, therefore, can we warrant relying on current knowledge to make decisions, particularly when the issues at stake are socially or politically sensitive, economically consequential, or deeply personal?[21]

In this chapter, I consider some examples in which scientists clearly went astray. The examples are drawn either from my own prior research and that of my students, or from historical examples that I have come to know well through three decades of teaching. Can we learn from these examples? Do they have anything in common? Might they help us answer the question of ex ante trust, by helping us to recognize cases where it may be appropriate to be skeptical, to reserve judgment, or to ask with good reason for more research?

I do not claim that these examples are representative, only that they are interesting and informative. They all come from the late nineteenth century onwards, because in my experience many scientists discount anything older on the grounds that we are smarter now, have better tools, or subject our claims to more comprehensive peer review.[22] Of course, no two historical cases are the same. Each of the examples I will present is complex, with more than one possible interpretation of how and why scientists took the positions they did. These cases do not define a "set." But they do have one crucial element in common: each of them includes red flags that were evident at the time.

Example 1:
The Limited Energy Theory

In 1873, Edward H. Clarke (1820–77), an American physician and Harvard Medical School professor, argued against the higher education of women on the grounds that it would adversely affect their fertility.[23] Specifically, he argued that the demands of higher education would cause their ovaries and uteri to shrink. In the words of Victorian scholars Elaine and English Showalter, "Higher education," Clarke believed, "was destroying the reproductive functions of American women by overworking them at a critical time in their physiological development."[24]

Clarke presented his conclusion as a hypothetic-deductive consequence of the theory of thermodynamics, specifically the first law: conservation of energy. Developed in the 1850s particularly by Rudolf Clausius, the first law of thermodynamics states that energy can be transformed or transferred but it cannot be created or destroyed. Therefore, the total amount of energy available in any closed system is constant. It stood to reason, Clarke argued, that activities that directed energy toward one organ or physiological system, such as the brain or nervous system, necessarily diverted it from another, such as the uterus or endocrine system. Clarke labeled his concept "The Limited Energy Theory."[25]

Scientists were inspired to consider the implications of thermodynamics in diverse domains, and Clarke's title might suggest he was applying energy conservation to a range of biological or medical questions.[26] But not so. For Clarke, the problem of limited energy was specifically female, i.e., female capacity. In his 1873 book, *Sex in Education; or, a Fair Chance for Girls*, Clarke applied the first law to argue that the body contained a finite amount of energy and therefore "energy consumed by one organ would be necessarily taken away from another."[27] But his was not a general theory of biology, it was a specific theory of reproduction. Reproduction, he (and others) believed, was unique, an "extraordinary task" requiring a "rapid expenditure of force."[28] The key claim, then, was that energy spent on studies would damage women's reproductive capacities. "A girl cannot spend more than four, or in occasional instances, five hours of force daily upon her studies" without risking damage, and once every four weeks she should have a complete rest from studies of any kind.[29] One might suppose that, on this theory, too much time or effort spent on any activity, including perhaps housework or child-rearing, might similarly affect women's fertility, but Dr. Clarke did not pursue that question. His concern was the potential effects of strenuous higher education.

In 1873, thermodynamics was a relatively new science, and Clarke presented his work as an exciting application of this important development. His book was widely read: *Sex in Education* enjoyed nineteen editions; over twelve thousand copies were printed in the three decades after its release. Historians have credited it with playing a significant role in undermining public support for educational and professional opportunities for women at that time; one contemporary commentator predicted that the book would "nip co-education in the bud."[30]

Clarke's argument was primarily aimed at co-education—that women could not withstand the rigors of a system of higher education designed for men—but it was also used against rigorous intellectual training for women of any sort, particularly that being conceptualized at the women's colleges that were being founded around that time, such as Smith (founded in 1871), Wellesley (1875), Radcliffe (1879), and Bryn Mawr (1885). Higher education for women was problematic, Clarke and his followers insisted, unless it was specifically designed to take account of women's "limited energy."[31] M. Carey Thomas, the first dean and second president of Bryn Mawr College, recalled that in the early years of the college, "we did not know when we began whether women's health could stand the strain of education." Early advocates of higher education for women were "haunted," she reflected, "by the clanging chains of that gloomy little specter, Dr. Edward H. Clarke's *Sex in Education*."[32]

Clarke's theory was also linked to emerging eugenic arguments (of which we will shortly say more). Like many elite white men in the late nineteenth and early twentieth centuries, Clarke feared the combination of women abandoning domestic responsibilities and the declining birth rate among native-born white women would be disastrous to the existing social order. He spoke for many when he fearfully predicted that "the race will be propagated from its inferior classes," and exhorted readers to "secure the survival and propagation of the fittest" by keeping women home, uneducated and child-rearing.[33] Perhaps for this reason his work was heralded by many male medical colleagues, who often shared these fears. One of these was Dr. Oliver Wendell Holmes, dean of the Harvard Medical School (and father of the future Supreme Court justice, who later defended the legality of eugenic sterilization in the infamous case of *Buck v. Bell*).[34] Holmes publicly expressed his "hearty concurrence with the views of Doctor Clarke."[35]

Clarke offered seven cases of young women who pursued traditionally male educational or work environments and experienced a variety of disorders, from menstrual pain and headaches to mental illness. His prescription to these women—and therefore to women in general—was to refrain from mental and physical effort, particularly during and after menstruation. Clarke did not attempt to measure or quantify the energy transfer among the body's organs, nor did he theorize the mechanism by which energy was selectively distributed to some parts of the body rather than others.[36] Rather, he asserted that his conclusion was a "deductive consequence from general scientific principles [i.e., the first law] using auxiliary assumptions." In this sense, his approach was similar to others at that time, such as social Darwinists, who also attempted to apply theories developed in the biological domain to problems in social worlds.

In hindsight it does not take much effort to identify the ways in which Clarke embedded prevailing gender prejudice and racial anxiety into his theory. But that risks historical anachronism. If our concern is how to identify problematic science, not in hindsight, but in our own time, then we must ask the question: Did anyone at the time object? The answer is yes. Feminists in the late nineteenth century found Clarke's agenda transparent and his non-empirical methodology ripe for attack. His leading critic within the medical community was Dr. Mary Putnam Jacobi, a professor of medicine at Columbia and the author of over a hundred medical papers.

Jacobi signposted the gender politics inside Clarke's theory, writing that the popularity of his work could be attributed to "many interests besides those of scientific truth. The public cares little about science, except insofar as its conclusions can be made to intervene in behalf of some moral, religious or social controversy."[37] She also identified its empirical inadequacy, based as it was on only seven women. As we saw in chapter 1, drawing deductive consequences from theory is part of accepted scientific methodology, but only part: deductive consequences have to be tested by reference to empirical evidence. And Clarke, Jacobi noted, didn't have much.

In 1877 she published a study of her own, *The Question of Rest for Women during Menstruation*, in which she sampled 268 women "who ranged in health, and education and professional status." (She also allowed the women to self-report their status, in contrast to Clarke who used his own interpretations of their symptoms.) Jacobi presented her data in a series of thirty-four tables examining the relationship between multiple variables, such as rest, exercise, and education.[38] She found 59% of women reported no suffering or only slight or occasional suffering from menstruation. Physiologically, she noted that there was "nothing in the nature of menstruation to imply the necessity, or even the desirability, of rest," particularly when the women's diets were normal. She supported this conclusion with a thorough literature review on menstruation and nutrition, as well as laboratory experiments on nutrition and the menstrual cycle.[39] Her research earned Harvard's Boylston Medical Prize. But it had little effect on Clarke or his male medical colleagues. In 1907 Dr. G. Stanley Hall wrote in his widely read work *Adolescence,* "it is, to say the very least, not yet proven that higher education of women is not injurious to their health."[40] Clarke's theory was viewed as sufficiently established as to place the burden of proof on those who claimed that higher education for women was fine.[41]

Example 2:
The Rejection of Continental Drift

In the 1920s and '30s, American earth scientists rejected a claim that forty years later was accepted as fact.[42] This was the claim that the continents were not fixed, but moved horizontally across the surface of the Earth; that these movements explained many aspects of geological history; and that the interactions of moving continents explained crucial geological features, such as the distribution of volcanoes and earthquakes. This concept came to be known as continental drift. Alfred Wegener, the prominent and respected geophysicist who proposed it, compiled a large body of empirical evidence drawn from existing geological literature.

While continental drift was not accepted at the time, there was broad consensus that existing theories were inadequate and an alternative explanation of the facts of geological history was needed. When the reality of drifting continents was accepted in the 1960s, in part based on these facts (as well as new ones that had come to light), some scientists were embarrassed to acknowledge that, not very long before, their community had rejected continental drift. In response, some suggested that continental drift had been rejected in the 1920s for lack of a mechanism to explain it. This was a plausible notion, and it was enshrined in many textbooks and even repeated by some historians and philosophers of science.[43] But it wasn't true. Several credible mechanisms had been offered at the time. None of these mechanisms was flawless—newly introduced theories rarely are—but scientists at the time had vigorous discussions about them, and some thought the mechanism issue had been resolved. American geologist Chester Longwell, for example, wrote that a model involving convection currents in the mantle—an idea that in the 1960s would be accepted as part of plate tectonics—was "a beautiful theory" that would be "epoch-making."[44]

If geologists had plausible mechanisms to explain continental drift, including ones that were later accepted, then why did they reject the theory? A telling element in this story was that American geologists were far more hostile to the theory than their European or British colleagues. Many continental Europeans accepted that pieces of the Earth's crust had moved over substantial horizontal distances; this was evident in the Swiss

Alps. Some British geologists also cautiously entertained the theory; in the 1950s and '60s many British school children learned about continental drift in their O- and A-level geology courses. But this was not the case in the United States: American geologists did not just reject the idea, they accused Wegener of *bad* science. This offers a rare opportunity to explore how scientists decide what constitutes good or bad science.

In debates over the theory, many American geologists explicitly raised methodological objections. In particular, they objected to the fact that Wegener had presented his theory in hypothetico-deductive form, which they considered to be a form of bias. Good science, they held, was inductive. Observation should precede theory and not the other way around. Edward Berry, a paleontologist at Johns Hopkins University, put it this way:

> My principal objection to the Wegener hypothesis rests on the author's method. This, in my opinion, is not scientific, but takes the familiar course of an initial idea, a selective search through the literature for corroborative evidence, ignoring most of the facts that are opposed to the idea, and ending in a state of auto-intoxication in which the subjective idea comes to be considered an objective fact.

Bailey Willis, chairman of the geology department at Stanford University and president of the Seismological Society of America, felt the books were "written by an advocate rather than an impartial investigator." Joseph Singewald, a geology professor at Johns Hopkins University, claimed Wegener "set out to prove the theory . . . rather than to test it."[45] Harry Fielding Reid, a founder of modern seismology, argued that the proper method of (all) science was induction. In 1922, he wrote a review of the English translation of Wegener's *Origin of Continents and Oceans* in which he described continental drift as a member of a species of failed hypotheses based on hypothetico-deductive reasoning.

> There have been many attempts to deduce the characteristics of the Earth form a hypothesis, but they have all failed . . . [Continental drift] is another of the same type. . . . Science has developed by the painstaking comparison of observations and through close induction, by taking one short step backward to their cause; not by first guessing at the cause and then deducing the phenomena.[46]

It has sometimes been suggested that comments such as these reflect an American rejection of theory in general. But American geologists did not reject theory per se. Many of them were actively involved in theory development in other domains. But they did have particular ideas about how scientific theories should be developed and defended. Scientific theory, they believed, should be developed inductively and defended modestly.

American geologists were suspicious of theoretical systems that claimed universal applicability and of the individuals who expounded them. One example was the "Neptunist" school, developed in the eighteenth century by Abraham Werner, which held that geological strata could be understood as the evolving deposits of a gradually receding universal ocean.[47] For many American geologists, Neptunism epitomized the type of grandiosity, operating under an authoritarian leader, that Americans discerned throughout European science. On a trip to Europe, Bailey Willis met Pierre Termier, director of the French Cartographic Service, who was known for his theory of *grande nappes*—the concept that large portions of the European Alps could be understood as mega-folds, created when a portion of continental crust was displaced over great lateral distances. Willis lamented that Termier was "an *authority*," whose theory young geologists in France "cannot decline to accept."[48]

The tone with which Willis discussed Termier explains what might otherwise be perplexing in this case: Scientists are supposed to be authorities, but the concern here is that this can slide into arrogance and dogmatism. It can slide into intellectual *authoritarianism*; Termier's authoritarian status could make it difficult for others to question his theory. The spirit of critical inquiry would be suppressed and scientific progress would be impeded, because no one would feel free to challenge or improve upon the idea.

The American preference for inductive methodology was thus linked by its advocates to American political ideals of pluralism, egalitarianism, open-mindedness, and democracy. They believed that Termier's approach was *typically* European—that European science, like European culture, tended toward the anti-democratic. American geologists thus explicitly linked their inductive methodology to American democracy and culture, arguing that the inductive method was the appropriate one for America because it refused to grant a privileged position to any theory and therefore any theorist. Deduction was consistent with autocratic European ways of thinking and acting; induction was consistent with democratic American ways of thinking and acting. Their methodological preferences were grounded in their political ideals.

This anti-authoritarian attitude was foregrounded by scientists who propounded the "method of multiple working hypotheses." Popularized by the University of Chicago geologist Thomas Chrowder Chamberlin, the method was an explicit methodological prescription for geological fieldwork. According to it, the geologist should not go into the field to test a hypothesis, but should first observe, and then begin to formulate

explanations through a "prism of conceptual receptivity that re-
fracted multiple explanatory options."[49] That meant developing
a set of "working hypotheses" and keeping all of them in mind
as work progressed. Chamberlin compared this to being a good
parent, who should not allow any one child to become a favorite. A good scientist was fair and equitable to all his working hypotheses, just as a good father loved all his sons. (Chamberlin
did not discuss daughters.)

The method was also a useful reminder that in complex geological problems the idea of a single cause was often wrong: many
geological phenomena were the result of diverse processes working together. It was not a matter of either/or but rather both/
and; the method of multiple working hypotheses helped geologists to keep this in mind. Chamberlin thought that the bitter
divisiveness that had characterized many debates in nineteenth-
century geology had arisen because one side had fixed on one
cause and the other side on another, rather than accepting that
the right answer might be a bit of both.[50] Scientists should be
investigators, not advocates. Chamberlin encapsulated this
idea in a paper called *Investigation vs. Propagandism*.[51]

At the University of Chicago, Chamberlin designed the graduate curriculum in geology specifically to train students to be
"individual and independent, not [merely] following of previous lines of thought ending in a predetermined result"—his gloss
of the European method. He also warned against the British system of empiricism, which he believed was "not the proper control and utilization of theoretical effort but its suppression."[52]
(Chamberlin was thinking specifically of Charles Lyell's denun-
ciations of high theory.) The method of multiple working hypotheses was the via media between dogmatic theory and empiricist extremism that would help in the future to avoid divisive
battles and factionalism.

One might wonder if this was just talk, but geologists' field notebooks and classroom notes from the period show that the method of multiple working hypotheses was practiced. Observations were segregated from interpretation, and geologists frequently listed various possible interpretations that occurred to them. One example is Harvard geologist Reginald Daly, an early advocate of continental drift. His field notebooks show how he enacted Chamberlin's prescription: in these notebooks he would record his observations on the left side of his notebook and, on the facing page, list a variety of possible interpretations of them. Reading Daly's field notes, one is reminded of Richard Hofstadter's famous claim that in the United States "a preference for hard work [was considered] better and more practical than commitments to broad and divisive abstractions."[53] It was not that American scientists were opposed to abstraction; it was that they were seeking a nondivisive approach to it. In the 1940s when Harvard professor Marland Billings taught global tectonics, he offered his students for their consideration no less than nineteen different theories of mountain-making, declining to say in class which one he preferred.[54] In this context, we can understand why American geologists reacted negatively to Wegener's work: He presented continental drift as a grand, unifying theory with the available evidence taken as confirmatory. For Americans, this was bad scientific method. It was deductive, it was authoritative, and it violated the principle of multiple working hypotheses. It was exactly what they expected from a European who *wanted* to be an authority.[55]

Americans, however, had become dogmatic in their anti-dogmatism, because in rejecting Wegener's theory on *methodological* grounds, they dismissed a *substantial body of evidence that in other contexts they accepted as correct.* Many of Wegener's harshest critics acknowledged this point, as when Yale geologist

Charles Schuchert allowed that the super continent of Gond-
wana "was a fact" that he "had to get rid of."[56] (Schuchert's solution was the ad hoc theory of "land bridges" to account for the
paleontological evidence, but which failed to explain the cor-
respondences in stratigraphy, which others sedulously analyzed.)
In later years, geologists would acknowledge that the evidence
that Wegener had marshalled was substantively correct.

Example 3: Eugenics

The history of eugenics is far more complex than the two examples we have just examined, in part because it involved a wide
range of participants, many of whom were not scientists (including US president Teddy Roosevelt), and the values and motivations that informed it were extremely diverse. Perhaps for this
reason some historians have been reluctant to draw conclusions
from what nearly all agree is a troubling chapter in the history
of science. But it has been used explicitly by climate change deniers to claim that because scientists were once wrong about
eugenics, they may be wrong now about climate change.[57] For
this reason, I think the subject cannot be ignored, and because
of its complexity I grant it more space than the two examples we
have just considered.

As is widely known, many scientists in the early twentieth
century believed that genes controlled a wide range of pheno-
typic traits, including a long list of undesirable or questionable
behaviors and afflictions, including prostitution, alcoholism, un-
employment, mental illness, "feeble-mindedness," shiftlessness,
the tendency toward criminality, and even *thalassophilia* (love
of the sea) as indicated by the tendency to join the US Navy or
Merchant Marine. This viewpoint was the basis for the social

movement *eugenics*: a variety of social practices intended to improve the quality of the American (or English, German, Scandinavian, or New Zealand) people, practices that in hindsight most of us view with dismay, outrage, even horror. These practices were discussed either under the affirmative rubrics of "race betterment" and "improvement," or the negative rubrics of preventing "racial degeneration" and "race suicide."[58] The ultimate expression of these views in Nazi Germany is well known. Less well known is that in the United States, eugenic practices included the forced sterilization of tens of thousands of US citizens (and principally targeting the disabled), a practice upheld in the *Buck v. Bell* decision, wherein Supreme Court justice Oliver Wendell Holmes, Jr., upheld the rights of states to "protect" themselves from "vicious protoplasm."[59]

The plaintiff in *Buck v. Bell* was a young woman, Carrie Buck, who had been sterilized after giving birth after being raped. State experts in Virginia testified that Carrie, her mother, and her child were all "feeble-minded"; this was used to warrant Carrie's sterilization to ensure that no further offspring would be produced. Justice Holmes encapsulated the decision in his memorable conclusion: "Three generations of imbeciles are enough."[60] Eugenic sterilization laws in the United States helped to inspire comparable laws in Nazi Germany, used to sterilize mentally ill patients and others deemed to be a threat to German blood; after World War II eugenics was largely discredited because of its relation to Nazi ideology and practices.[61]

We might be tempted to dismiss eugenics as a political misuse of science, insofar as it was promoted and applied by people who were not scientists, such as President Roosevelt or Adolf Hitler, or by men who worked in eugenics but were not trained in genetics, such as the superintendent of the Eugenics Record Office, Harry Laughlin, who testified in the US Congress on behalf of eugenic-based immigration restrictions.[62] But that only gets us so far, insofar as eugenics was developed and promoted to a significant extent by biologists, and by researchers who came to be known as "eugenicists." Moreover, like Clarke's Limited Energy Theory, eugenics was presented as a deduction from accepted theory, in this case Charles Darwin's theory of evolution by natural selection. If, as Darwin argued, traits were passed down from parent to offspring, and fitness was increased by the differential reproduction and survival of fit individuals, then it stood to reason that the human race could be improved through conscious selection. Darwin had developed his theory of natural selection in part by observing selective breeding by pigeon fanciers: breeding was the deliberate and conscious selection of individuals with desirable traits to reproduce and the culling of those with undesirable ones. If breeders improved their pigeons, dogs, cattle, and sheep through selection, was it not obvious that the same should be done for humans? Should we not pay at least as much attention to the quality of our human offspring as of our sheep? And was it therefore not equally obvious that society should take steps to encourage the fit to reproduce and discourage the unfit? This latter question was famously posed by Thomas Malthus in the eighteenth century, who argued against forms of charity that might encourage the poor to have more children, and who, through his arguments about the inexorable mathematics of reproduction, inspired Darwin.[63]

The founder of "scientific" eugenics is generally taken to be Darwin's cousin Francis Galton (1822–1911) and many elements in Darwin's work seemed to support the view that the laws of selection that operate in nature must also operate in human society. In *The Descent of Man* (1871), for example, Darwin made clear that he believed that natural selection applied to men as well as beasts, and he argued that some human social practices, such as primogeniture, were maladaptive. It was not a stretch to read Darwin as suggesting that human laws and social practice should be adjusted to account for natural law.

For humans, Galton argued, the most important trait was intelligence, and so Galton undertook the study of intelligence and heredity. Many physical traits, such as height, hair, skin and eye color, and even overall appearance, seemed to be largely inherited, but was intelligence? In his 1892 work *Hereditary Genius*, Galton concluded that it was. Analyzing the family trees of "distinguished men" of Europe, he found that a disproportionate number came from wealthy or otherwise notable families. While he recognized that "distinction" was not the same as "intelligence," Galton used it as a proxy. Finding that distinctions of all types—political, economic, artistic—did cluster, he concluded that character traits ran in families just as physical ones did.[64]

Galton, however, observed one crucial difficulty, what he called the *law of reversion to the mediocre*: that the offspring of distinguished parents tended to be more mediocre—that is to say, more average—than the parents. He illustrated this with height: tall couples gave birth, on average, to children who were not as tall as they. In an early insight into what we would now call population genetics, Galton reasoned that the children were inheriting traits not only from their parents, but also their grandparents and great-grandparents, i.e., their entire family tree. The same would be true of any trait, including intelligence.

Galton's conclusions regarding the prospects for overall improvement of the human race were thus pessimistic, because if offspring inherited from their entire family tree then improvements would take many generations to achieve. Pigeon fanciers and dog breeders did not achieve their results in a single generation, but by patient selection over years and decades, and for human breeding this was unrealistic. Galton did suggest vaguely that "steps" should be taken to encourage the "best" to procreate more and the "worst" to procreate less so as "to improve the racial qualities of future generations" and avoid "racial degeneration."[65] Such non-coercive encouragement came to be known as "positive eugenics." But Galton was not optimistic that a program of eugenics could be readily or reasonably achieved. The law of reversion to the mediocre seemed to undermine such aspirations. Others, however, insisted not only that human improvement through breeding could be achieved, but that it needed to be.

Today, the idea of "racial degeneration" is impossible to separate from its Nazi associations, but in the early twentieth century the threat was keenly felt—at least by many white men— as real and present, and the eugenic ideal was taken up by physicians, scientists, intellectuals, and political leaders. In the United States, besides Teddy Roosevelt, another prominent eugenicist was the conservationist Madison Grant, a founder of the Save-the-Redwoods League, trustee of the American Museum of Natural History, and author of the popular book *The Passing of the Great Race* (1916).[66] This was the Nordic "race"— what we might now call white Anglo-Saxons—which Grant believed was threatened by the weaker "races" of Jews, southern Europeans, and Negros. These latter groups, he argued, should be isolated in ghettos and prevented from interbreeding with men and women of northern European descent. Grant's arguments played a role in the Johnson-Reed Immigration Restriction Act of 1924, which limited immigration from southern and eastern Europe to no more than 2% of the US population as measured in the 1890 census and completely eliminated immigration from Asia.[67] Stephen Jay Gould characterized *The Passing of the Great Race* as the most influential work of scientific racism ever published in America; historian Jonathan Spiro notes that it was widely embraced in Nazi Germany, including by Hitler, who wrote to Grant saying, "The book is my Bible."[68]

Eugenics as a social movement grew dramatically in the years 1910–20, with a proliferation of books and articles on race and fitness, nearly all of which were framed as applications of biological science. As Grant crisply put it, "The laws of nature require the obliteration of the unfit."[69] The rediscovery in 1900 by Hugo de Vries and colleagues of the work of Gregor Mendel, and the support it seemed to give—indeed, the proof, in some eyes— for the hard inheritance of characteristics, was important to the surge of support for eugenics, as Mendel's findings seemed to rule out Lamarckian notions that individual improvement could be effected via environmental improvement.[70]

In the United States, the locus of scientific eugenics was the Eugenics Record Office (ERO), founded in 1910 at Cold Spring Harbor, Long Island, and later incorporated as a department within the Carnegie Institution of Washington Station for Experimental Evolution.[71] Its director was Charles Davenport, a professor of biology at the University of Chicago and pioneer in biometrics. In founding the ERO, Davenport declared in language that Oliver Wendell Holmes, Jr., would echo, "Society needs to protect itself; as it claims the right to deprive the murderer of his life so also it may annihilate the hideous serpent of hopelessly vicious protoplasm."[72]

One could not do experiments on humans as Mendel did on peas, but one could collect data, and Davenport launched a major study on "heredity in relation to eugenics."[73] The goal was to establish the scientific basis of human inheritance through study of family histories; the methodology was to hire field workers to interview families about their histories. (In this regard, the activities at the ERO were quite different from the work of the biologists at the adjacent experimental station.) Trained field workers asked questions about such behaviors as alcoholism, prostitution, gambling, promiscuity, and criminality; physical "defects" including hermaphrodism, cleft palate, and polydactyly; illnesses such as hemophilia and tuberculosis; mental "defect" such as "feeblemindedness," schizophrenia, and other forms of mental illness; and the general category of social attainment and accomplishment.

Between 1911 and 1924, 250 field workers, mostly women, trained at the ERO and were sent out to collect these data. The answers were recorded on index cards. The field workers found that these traits often did run in families. Davenport therefore concluded that social remedies were needed to prevent reproduction by parents carrying "undesirable" trains, and he became an advocate of "segregation"—to keep the mentally and physically ill in home and asylums where they could not breed—and sterilization—to ensure that the unfit, both incarcerated and at large—would not reproduce.

His deputy, Harry Laughlin, used the ERO results to promulgate "Model Sterilization Laws," and to testify in Congress to the desirability of restricting immigration from southern and central Europe. ERO data demonstrated, he claimed, that immigrants were more likely than native-born Americans to commit crimes, and that this tendency toward criminality was inherited. In 1924, the US Congress passed the Johnson-Reed Act, which severely restricted immigration along eugenic lines.[74]

In the 1930s, thirty-two states in the Union passed sterilization laws, and at least thirty thousand US citizens were sterilized, mostly without informed consent and sometimes without their knowledge.[75]

Laughlin was a hero to many Nazis. In 1936, he was awarded an honorary degree from the University of Heidelberg for his work on the "science of racial cleansing." It has been said that the Nazis based their own sterilization laws on the model laws developed by Laughlin at ERO.[76] At Nuremberg, one profferd defense was that Nazi laws were based on what Americans had advocated.

As I have already noted, eugenics was complicated. Historian Daniel Kevles has argued that eugenics had several principal components, intermixed in various ways:[77]

- *Social control* of reproduction, either through control of marriage or isolation in asylums, jails, and other institutions;
- *Natalism.* Encouraging large families among the "fit" (generally understood to be wealthy and white) and discouraging of reproduction among the "unfit" (everyone else);
- *Malthusianism.* Discouraging social welfare programs, including universal education, minimum wage laws, and public health measures intended to reduce infant mortality on the grounds that they ran against the natural laws that would otherwise weed out the unfit. Eugenicists also discouraged birth control, assuming that those who should use it would not and those who shouldn't use it would;
- *Hereditiarianism and anti-environmentalism.* Rejecting the role of environment and locating the cause of social position and behavioral traits exclusively or nearly exclusively in physical inheritance; and
- *Racial anxiety.* Fearing that breeding of the unfit, coupled with immigration, was polluting or diluting the racial identity of the country, leading to "national" or "racial" deterioration, with those two terms and concepts often used interchangeably.

To this list we may add

- *Gender anxiety.* Eugenic arguments were often coupled to arguments against women's participation in the work force and the promotion of a constricted role centered on home and family.[78]

Most of these elements—racial anxiety, gender anxiety, natalism—are not scientific values, which raises the question: What exactly was the role of science and scientists in the eugenics movement?

It is sometimes claimed that there was a scientific consensus supporting eugenics, and therefore we are justified in disbelieving or rejecting contemporary matters about which there is a scientific consensus. The novelist Michael Crichton, for example, used this argument to try to discredit climate science, likening contemporary calls for action to prevent anthropogenic climate change to earlier calls to prevent race suicide.[79] Both, he suggested, were politics masquerading as science.

The fact that scientists may have been wrong about some matter in the past in no way tells us whether they are right or wrong about some wholly unrelated matter today, but Crichton's argument does remind us that scientists have not always been on the side of the angels. Insofar as eugenics, like the Limited Energy Theory, was conceptualized and justified as a logical deduction from scientific theory, we cannot simply explain it away as a "misuse" or "misapplication" of science. So was there a scientific consensus on eugenics? The short answer is no.[80] Prominent social scientists and geneticists objected to eugenic claims. As historian Garland Allen has put it, "It was not the case that nearly everyone in the early twentieth century accepted eugenic conclusions."[81]

Social scientists made a complaint that is obvious in retrospect and often invoked today in nature-nurture debates: that many of the ills recorded by field workers could be explained by bad nutrition, bad education, lack of linguistic skills, and/or bad luck. It was possible that genetics explained adverse outcomes, but so could many other things. The observation of an adverse outcome was no proof of the genetic theory of its causation.

Many poor whites in the 1910s and '20s were immigrants who faced numerous obstacles, including overt discrimination in employment and lack of adequate health care. Reformers pointed to the many immigrant children who had "improved themselves" with the help of education and other social programs, demonstrating that social reforms if seriously pursued could work to improve outcomes. The German-Jewish immigrant anthropologist Franz Boas, in particular, argued that while traits like hair and eye color might be wholly inherited—and there was scientific evidence from laboratory and breeding studies to suggest that this was so—other matters were not easily so reduced. Height, one of Francis Galton's favorite topics of study, was a case in point. A person's stature, Boas remarked, is partly inherited, but "is also greatly influenced by more or less favorable conditions during the period of growth."[82] Insufficient science had been done to understand the interplay between physical and social factors in determining developmental outcomes, and in the absence of understanding this interplay it was wrong to assume that any complex trait was controlled by genetics, and certainly wrong to assume that it was wholly so.

Boas particularly objected to claims regarding the hereditary character of intelligence. IQ tests had not been shown to measure anything meaningful, and there was no evidence of racially specific hereditary mental or behavioral traits in blacks, immigrants, or any other group.[83] We could observe disparate outcomes, but we had no independent evidence of the causes of those outcomes. On the contrary, there was evidence of social causes: Boas's student Margaret Mead had shown in her 1924 master's thesis that the scores on IQ tests of the children of Italian immigrants varied according to family social status, length of time in the United States, and whether English was spoken at home.[84]

Mead's discussion of Italian immigrants is an important reminder that, while the language of eugenics was that of "racial degeneration," eugenics in America was concerned both with issues of race (as we understand the term today) and with gradations of European ethnicity, both of which were tied to class.[85] The threat was understood to be to the "Nordic race"—the peoples of northern European descent—from both European and non-European sources, and so a major focus of eugenic study and target of eugenic practice was poor whites. In the United States, that largely meant immigrants, but in the United Kingdom it meant the working class. For this reason, it is perhaps not surprising that another group of scientists who objected to eugenics were socialists, including the British geneticists J.B.S. Haldane, J. D. Bernal, and Julian Huxley, and the American socialist Herman Muller.[86]

Professor of genetics and biometry at University College London, J.B.S. Haldane was the son of the famed Oxford physiologist John Scott Haldane, a socialist who pioneered the study of occupational hazards and originated the practice of bringing canaries into coalmines to monitor air quality.[87] Initially Haldane sympathized with aspects of eugenics—in college he joined the Oxford Eugenics Society—but he was soon offended by its evident sociopolitical prejudices, particularly its class bias.

Haldane highlighted the thin empirical basis for eugenic claims, particularly given that the mechanisms of inheritance, particularly of complex traits, were only just now coming into scientific focus. Too little was known about heredity to justify any eugenics program, and "many of the deeds done in America in the name of eugenics are about as much justified by science as were the proceedings of the inquisition by the gospels." He opposed all sterilization programs, including voluntary ones, on the grounds that "any legislation which does not purport to apply, and is not actually applied (a very different thing) to all social classes alike, will probably be unjustly applied to the poor." He also insisted in the value and dignity of the working class: "A man who can look after pigs or do any other steady work has a value to society and . . . we have no right whatever to prevent him from reproducing his like."[88]

Haldane did not believe that "the theory of absolute racial equality" was necessarily correct, but he thought that any actual difference—either of type of degree—would be difficult to establish objectively. The best one might hope for would be to establish difference between populations—as Galton had done—but that would not tell you anything meaningful about the characteristics, much less the social value, of any individual. Perhaps reflecting on his acquaintance with the great American actor and singer Paul Robeson, he insisted that "it is quite certain that some negroes are intellectually superior to most Englishmen."[89]

Herman Muller, who shared the Nobel Prize for his work demonstrating that x-rays could induce heritable genetic changes in fruit flies, also objected to eugenics. Muller is a complex case, insofar as he did not doubt that the human race could in principle be improved through eugenic practices. Nor did he doubt that ideally it should be. But equally firmly he believed that improvement would never happen equitably under capitalism.

Muller was the principal author of the 1939 *Geneticists' Manifesto*, signed by twenty-two American and British scientists (as well as the historian of science Joseph Needham), in response to a request from the Science Service to answer the query, "How could the world's population be improved most effectively genetically?"[90] Muller and his colleagues rejected the premise that this question could be answered biologically. They began by insisting that the question "raises far broader problems than the purely biological ones, problems which the biologist unavoidably encounters as soon as he tries to get the principles of his own special field into practice."[91] In other words, this was not only or even primarily a biological question:

> For the effective genetic improvement of mankind is dependent upon major changes in social conditions, and correlative changes in human attitudes. In the first place, there can be no valid basis for estimating and comparing the intrinsic worth of different individuals, without economic and social conditions which provide approximately equal opportunities for all members of society instead of stratifying them from birth into classes with widely different privileges.[92]

These men were not unilaterally opposed to efforts to make genetic improvements to the human race. Even after revelations of Nazi atrocities, Muller continued to support the idea of deliberate human improvement, arguing in 1954 that "the fact that the so-called eugenics of the past was so mistaken . . . is no more argument against eugenics as a general proposition than say the failure of democracy in ancient Greece is a valid argument against democracy in general."[93] (This would have been an interesting response to Michael Crichton.) But Muller and his colleagues rejected much if not all of the evidence being invoked to support eugenic claims, because reigning accounts assumed a level social playing field that patently did not exist.

Existing studies assumed that observed differences were genetic: in effect assuming the thing they were intended to prove.

The authors of *Geneticists' Manifesto* accepted that there were both genetic and environmental aspects of intelligence, behavior, social accomplishments, and many other things—as indeed most scientists accept today. But existing studies failed to identify the relative contributions of social and genetic elements in human characteristics.

> Before people in general, or the state which is supposed to represent them, can be relied upon to adopt rational policies for the guidance of their reproduction, there will have to be ... a far wider spread of knowledge of biological principles and recognition of the truth that both environment and heredity constitute dominating and inescapable complementary factors in human well-being.[94]

No real advance could be made, they held, without "the removal of race prejudices and of the unscientific doctrine that good or bad genes are the monopoly of particular peoples or of persons with features of a given kind," and this would not occur until "the conditions which make for war and economic exploitation have been eliminated [through] some effective sort of federation of the whole world based on the common interests of its peoples." Capitalist societies manifestly did not provide "approximately equal opportunities for all." Eugenics could not work under capitalism: the lower classes would always be targeted.

A level playing field would only be a start, moreover, because it was unreasonable to expect any parent to worry about the state of future generations unless they were first "extended adequate economic, medical, educational and other aids in the bearing and rearing of children." It was also unreasonable to expect intelligent women to abandon their personal interests and aspirations on behalf of improving the population at large; the scientists therefore suggested the need for social policies to ensure that a woman's "reproductive duties do not interfere too greatly with her opportunities to participate in the life and work of the community at large." This meant that workplaces needed to be "adapted to the needs of parents and especially mothers," and that towns and community services needed to be reshaped "with the good of children as one of their main objectives." It also meant that women needed access to safe and effective birth control: "A . . . prerequisite for effective genetic improvement is the legalization, the universal dissemination, and the further development through scientific investigation, of ever more efficacious means of birth control . . . that can be put into effect at all stages of the reproductive process," including voluntary sterilization and abortion.[95]

Finally, these geneticists noted, to improve the world through selection would require agreement about what constituted improvement, something that was by no means apparent, particularly if the goals of selection were social ones. In their view, the most important genetic characteristics one might want to try to foster would be those for health, for the "complex called intelligence," and "for those temperamental qualities which favour fellow-feeling and social behavior, rather than those (to-day most esteemed by many) which make for personal 'success,' as success is usually understood at present."[96] So despite expressing in-principle support for eugenic ideals, they opposed eugenic proposals in practice. The prerequisite for improving the quality of the world's population was improving the social conditions of the world.[97]

The socialist geneticists' opposition to eugenics was rooted in their politics, but one did not have to be a socialist (or social scientist) to recognize flaws in eugenic research. In particular, many geneticists pointed out the fallacy of conflating genes with outcomes. Garland Allen has stressed that the great British statistician Karl Pearson, who was a eugenicist, strongly criticized the work of the ERO as "carelessly and sloppily conceived and executed, and lack[ing] any semblance of normal scientific rigor."[98]

Herbert Spencer Jennings was an American geneticist at Johns Hopkins University known for his 1930 book, *The Biological Basis of Human Nature*.[99] While the title might suggest an argument for genetic determinism, the book presented the scientific case for the interaction of genes and environment. Against the genetic determinists, Jennings wrote:

> A given civilization is the outgrowth of the interaction of the genetic constitutions present in the population, with the environment—including knowledge, inventions, traditions—of that population. By changes in the latter set of factors enormous differences have in the past been made in the cultural system. . . . No cultural system is the outgrowth of genetic constitution alone.[100]

And against the environmental determinists:

> [The environmental determinist argues that] by subjection to adequately diverse environments, diverse training and instruction, any of [a group of people] can be made . . . into "doctor, lawyer, merchant, chief" . . . Biology has no proper quarrel with such an assertion. What an enlightened view of biology would add . . . is this: While any of the normal individuals, taken early and properly guided, could be made into physicians, it would take different treatment to accomplish that end in the different individuals.[101]

Eugenicists had committed numerous logical and methodological fallacies, including being overly influenced by implicit assumptions ("underlying . . . but never stated"), ignoring evidence that did not support their positions, and persisting in "mistaken conclusions after the discovery that they are mistakes."[102] Jennings was particularly critical of what he called "the fallacy of non-experimental judgments," noting that precisely because nearly everyone had an opinion on heredity and evolution, it was "essential to set aside prior views and build one's opinions on the basis of experimental evidence." But this was just what most eugenicists failed to do; they ran with their priors and ignored disconfirming evidence.[103] Jennings also noted the widespread use of what today we would call the fallacy of the excluded middle—to assume that because some traits have been shown to be inherited, all traits are inherited, and vice versa regarding the environment—and, in language reminiscent of T. C. Chamberlin, "The fallacy of attributing to one cause what is due to many causes."[104]

For Jennings it was obvious that the answer to the nature/nurture debate was both/and. He made the point in a 1924 article by analogy to material objects:

> What happens in any object—a piece of steel, a piece of ice, a machine, an organism—depends on the one hand upon the material of which it is composed [and] on the other hand upon the conditions in which it is found. Under the same conditions objects of different material behave diversely; under diverse conditions objects of the same material behave diversely. . . . Neither the material constitution alone, nor the conditions alone, will account for any event whatever; it is always the combination that has to be considered.

And so it was for organisms. "The individual is produced by the interaction of genes and environmental conditions; so that the same set of genes may yield diverse characteristics under diverse environments." Eugenics was doomed to fail, because "behavior is bound to be relative to environment, it cannot be dealt with as dependent on genes alone. A given set of genes may result under one environment in criminality; under another in the career of a useful citizen."[105]

Jennings is but one example; if space permitted we could easily multiply his critique. The Nobel Laureate T. H. Morgan, famous for his work on the genetics of fruit flies, stressed in the 1920s that the problems eugenicists proposed to repair would likely be more quickly remedied through social reform than through selective breeding.[106] Many non-scientists also raised methodological and moral objections.[107] (And there were objections raised in other countries that I have not considered here.)[108] The important point here is that eugenics as a political movement in important ways conflicted with scientific understanding, and it is simply not correct to say that there was a scientific consensus on eugenics.[109]

Now let us consider an example where there was a consensus, but one that ignored or at least discounted important, significant evidence.

Example 4: Hormonal Birth Control and Depression

Many women have had the experience of becoming depressed or melancholic on taking the contraceptive pill, many doctors are aware of their patients' experience, and many scientific studies have affirmed this link. Indeed, some of the earliest studies of the effects of the Pill in the late 1950s noted side effects including "crying spells" and "irritability," and the package insert that now comes with it states that one of the known side effects is "mental depression" (see fig. 1).

ADVERSE REACTIONS

An increased risk of the following serious adverse reactions has been associated with the use of oral contraceptives (see WARNINGS section).

- Thrombophlebitis and venous thrombosis with or without embolism
- Arterial thromboembolism
- Pulmonary embolism
- Myocardial infarction
- Cerebral hemorrhage
- Cerebral thrombosis
- Hypertension
- Gallbladder disease
- Hepatic adenomas or benign liver tumors

There is evidence of an association between the following conditions and the use of oral contraceptives:

- Mesenteric thrombosis
- Retinal thrombosis

The following adverse reactions have been reported in patients receiving oral contraceptives and are believed to be drug-related:

- Nausea
- Vomiting
- Gastrointestinal symptoms (such as abdominal cramps and bloating)
- Breakthrough bleeding
- Spotting
- Change in menstrual flow
- Amenorrhea
- Temporary infertility after discontinuation of treatment
- Edema
- Melasma which may persist
- Breast changes: tenderness, enlargement, secretion
- Change in weight (increase or decrease)
- Change in cervical erosion and secretion
- Diminution in lactation when given immediately postpartum
- Cholestatic jaundice
- Migraine
- Allergic reaction, including rash, urticaria, and angioedema
- Mental depression
- Reduced tolerance to carbohydrates
- Vaginal candidiasis
- Change in corneal curvature (steepening)
- Intolerance to contact lenses

FIGURE 1. Detail of package insert for the oral, hormonal contraceptive ORTHO TRI-CYCLEN® Lo Tablets (norgestimate/ethinyl estradiol) showing "mental depression" among the list of potential adverse reactions, which are "believed to be drug-related."

Recently, there was a flurry of media attention about a new study demonstrating that the Pill can cause depression.[110] Physicians lauded the study, and the media presented the result as a novel finding.[111] My own daughter, however, asked me on the day the coverage hit the media: How is this *news*? She knew that the Pill could cause depression, because I had told her so.

I have no history of depression—no family history of depression or mental illness of any sort—but when I was in my midtwenties, I experienced a sudden and peculiar bout of extreme melancholy. I lost my energy for daily tasks, lost interest in my work, and, after about six weeks, found myself having trouble getting out of bed. And yet, in other respects my life was going well. I was in my second year of graduate school, had done very well in my first year, was working on an exciting project for which I had adequate funding, and had met a very nice man who would soon become my husband (and to whom I've now been married for more than thirty years).

I went to counseling at a campus health center, and I was lucky. The female counselor asked me straight away: Are you on the Pill? The answer was yes. I explained that I had recently returned from Australia, and because Australia at that time had free health insurance, including prescription drugs, I had bought a year's worth before I left. But the particular formulation that I had been prescribed in Australia was not available in the United States, so when the year was up I had to switch to another type. That had occurred two months before. The onset of my depression began shortly after I had started this new form of the Pill. The therapist told me that the type of pill I was now on—a combination formulation—was well known to be more likely to cause depression than some other options. I stopped the drug immediately and my recovery began nearly as immediately. Within a few weeks I was back to my normal self, I thanked the therapist, and went on to a successful academic career and life.

My experience can be dismissed as "just an anecdote," but I prefer to view it as a clinical study in which n = 1. The more important point is that many women have had such experiences and reported them to their physicians and therapists. The website Healthline.com, which claims to be the "fastest growing consumer health information site," notes that "depression is the most common reason women stop using birth control pills."[112] Moreover, like me, many women have bounced back to normal when they stopped taking the Pill or switched to other formulations. And these case reports have spurred numerous scientific studies. As one physician recently wrote, "decades of reports of mood changes associated with these hormone medications have spurred multiple research studies." So my daughter was correct to ask: how was this new study news?

One answer was offered by Monique Tello, a practicing MD, MPH, who writes for the *Harvard Gazette*: "The study of over a million Danish women over age 14, using hard data like diagnosis codes and prescription records, strongly suggests that there is an increased risk of depression associated with *all* types of hormonal contraception." Previous studies, in contrast, were all "of poor quality, relying on iffy methods like self-reporting, recall, and insufficient numbers of subjects." The authors of the new study concluded that previously it had been "impossible to draw any firm conclusions from the research on this subject."[113]

It is hard to argue with a study of over one million women. It is also hard to argue with any study done in Denmark, which has a national health care database covering every Danish citizen and thus allows researchers to correct for sampling biases and other confounding effects. It is thanks to Denmark that we can say with confidence that children who are fully vaccinated according to prevailing public health recommendations do not suffer autism at greater rates than those who are not.[114] So, three cheers for Denmark. Three cheers, as well, for this big, new convincing study. But note the explanation of why it took so long to come to this point: the lack of "hard data like diagnosis codes and prescription records." Previous studies, we are told, relied on "iffy methods like self-reporting, recall, and insufficient numbers of subjects."[115]

The term "hard data" should be a red flag, because the history and sociology of science show that there are no hard data. Facts are "hardened" through persuasion and their use. Moreover, remarks of this type raise the question of why some forms of data are considered hard and others are not. Just look at what is being considered hard data here: diagnosis codes and prescription records. Many people would say hard data are quantitative data, but neither of these constitutes a measurement: they are the subjective judgments of practitioners and the drugs they choose to prescribe in response to those judgments.[116] Moreover, there is a substantial literature on misdiagnosis in medicine, and on the distorting effects of pharmaceutical industry advertising and marketing on prescribing practices.[117] Given what we know about medical practice and its history, the idea that diagnosis codes and prescription records should be taken as hard facts seems almost satirical.

But it gets worse: the study authors accepted the reports of doctors—their diagnosis codes and prescription records—as facts, whereas the reports of female patients were dismissed unreliable—in Tello's words: "iffy." Bias—either against women or against patients—is clearly on display. But here is the key point: the conclusion of the Denmark study is the *same* as all those iffy, self-reports from female patients. If the new study is correct, then the allegedly iffy self-reports were correct all along.

These self-reports involved millions of women, too. The Pill has been on the market in the United States and Europe since the early 1960s. According to the CDC, during period 2006–10, over ten million American women took the Pill.[118] According to the World Health Organization, over one hundred million are currently taking it worldwide.[119] While self-reporting does not offer a good basis for an accurate quantitative assessment of risk of depression caused by hormonal contraception, it surely offers important qualitative evidence. It seems extremely unlikely that all the women who reported mood changes while on the Pill were simply confused or making it up.

In fact, the connection between hormonal birth control and depression has been known almost as long as the Pill has been on the market. In 1969, feminist journalist Barbara Seaman published *The Doctor's Case against the Pill*, a book that helped to launch the women's health movement. Seaman's book made women and doctors around the country aware of the serious health risks of the Pill as it was then formulated, and led to congressional hearings resulting in the first package insert to warn against risks involved with a prescription medication. Chapter 15 of her book was entitled "Depression and the Pill," and it began:

> Psychiatrists were among the first doctors to persuade their own wives to stop using birth control pills. Finely tuned to emotional feedback, they did not take long to notice certain adverse reactions in their wives and daughters, patients and friends. The effects that were the most obvious ranged from suicidal and even murderous tendencies to increased irritability and tearfulness. . . . A few pill-users have become so hostile, suspicious and delusional that they have seriously thought of murdering—or have actually attempted to murder—their own husbands and children. Others attempt to commit suicide and some have succeeded.[120]

Within a few years of the Pill coming on the market, adverse mental health effects had been widely reported. A 1968 study in the United Kingdom looked at 797 women who took oral contraceptives; many reported emotional side effects and two committed suicide.[121] By 1969 British researchers had found that one in three Pill users experienced personality changes; three in fifty who were studied became suicidal. A US study by researchers at the University of North Carolina School of Medicine found that 34% of otherwise healthy young women reported depression after starting on the Pill. These studies did not include control groups, but one in Sweden compared two groups of postpartum women, matched with respect to social background, previous history of depression, and other factors. It found significantly higher rates of psychiatric symptoms in women who went on the Pill after giving birth than in the group who used other forms of birth control.

We cannot judge from Seaman's account how good any of these studies were; her point was that to the extent that scientists had examined the question, they had found evidence to support women's accounts, accounts that formed the emotional heart of Seaman's story. She told of women who became agitated and disorganized; who experienced panic attacks in movie theatres; who set "accidental" fires; who found themselves weeping uncontrollably for no apparent reason; and who felt themselves to be on the verge of breakdown. Some of these women may have been depressed for other reasons, but Seaman supported their accounts with testimony from psychiatrists. Women's stories formed the emotional center of the book, but doctor's stories provided the intellectual center. It was not the *patient's* case against the Pill, but the *doctor's* case.

With respect to the mental health effects, the key doctors were psychiatrists, to whom women had gone for help after becoming depressed on the Pill, or who noticed changes in their wives and friends and in patients they had been seeing for some time and knew well. Seaman quoted one Manhattan psychiatrist describing the resistance he initially encountered from other doctors:

> My fights with the gynecologists began in 1963 [three years after Enovid, the first oral contraceptive, was approved.][122] I'd been seeing one patient twice a week for two years. . . . She was tough as nails . . . Her father had been an alcoholic. She'd fought her way to the top as a fashion model. . . . She's one of the most sensible patients I ever had. Exploitative? Yes. Neurotic? A little. Depressed? Never. Eight days after this patient went on the pill, she arrived for her appointment and wept through the whole session. The same thing happened the next time and the next . . . She talked about "giving up" and "ending it all." I suggested that she get off the pill. We'd see what happened then. She did. The next time I saw her she was her old self. But then came the first in a series of calls from her gynecologist. In essence what he had to say was, "You stick to your own unraveling or whatever it is you do, and let me take care of *my* knitting. Birth control is not a psychiatrists' province."[123]

(Eventually, gynecologists would accept there was a pattern; this particular gynecologist was convinced by the psychiatrist and began to send him patients for Pill-induced depression.)

Other psychiatrists told similar stories. Patients they had known for years were suddenly different; or patients were sent by their families because of sudden, frightening changes. The Atlanta physician John R. McCain presented a paper at the New England Obstetrical and Gynecological Society warning that the mental health effects of the Pill were "among the complications which seem to have the most serious potential danger."[124] The good news, many doctors noted, was that when women went off the Pill, the abatement of symptoms was as fast as the onset. This, of course, was further evidence that the Pill was a factor in their condition.

In many of these stories, women noted that their mood swings and depression were similar to what they had experienced when they were pregnant or just after giving birth—and doctors had rarely doubted that hormones had something to do with those experiences! Among the various stories recounted in the book, my personal favorite is this one: "When I was on the pill," one psychiatrist's wife reported, "I hardly ever got off the couch except to slap one of the children."[125]

In the years that followed, scientists and physicians undertook studies of the mental health effects of the Pill. But considering how many women have taken the Pill, the total number of studies is startling modest. A quick PubMed search in 2016 on hormonal birth control and depression/mood or psychological disorders/libido changes found twenty-seven papers. This may be an underestimate—other key words or phrases might have turned up more, and mood changes may also have been detected in studies concentrating on other things—but compare this to another issue that I have studied: climate change. In my 2004 study of climate science, I used a sample of just under one thousand articles to estimate the state of scientific opinion.[126] That sample came from a population that was estimated to be over ten thousand papers. Since that time, at least that many more have been published.[127] Given that over one hundred million women are on the Pill today, doesn't it seem troubling that there are so few studies on something that was recognized as a potentially serious problem more than fifty years ago?

Mood changes are admittedly a difficult thing to study and almost impossible to quantify. Feelings are, by definition, subjective, and depression cannot be measured in the sort of way that cholesterol or high blood pressure can be. But consider this: in 2016, a clinical trial of a male hormonal contraceptive injection in 320 men was *abandoned* after the men taking part reported increased incidences of adverse effects, including changes in libido and mood disorders. In fact, more than 20% reported mood disorders. One man developed severe depression; another tried to commit suicide. Because of the adverse effects, the trial was halted—even though the rate of pregnancy suppression was more than 98%. The researchers reported:

> The study regimen led to near-complete and reversible suppression of spermatogenesis. The contraceptive efficacy was relatively good compared with other reversible methods available for men. The frequencies of mild to moderate mood disorders were relatively high.[128]

The male hormonal contraceptive injection was shown to work as well as the Pill, yet the clinical trial was stopped because of adverse effects, one of which was a dramatic increase in mood disorders.[129] If you are wondering how the researchers measured this, the answer is: self-reporting.

This result could have been predicted, not only because similar effects were seen in women, but because there is a mechanism that explains why hormonal contraceptives have this effect. It is the link between reproductive hormones and serotonin.

> Low levels of serotonin, a neurotransmitter in the brain, have been linked to depression. High levels of estrogen, as in first-generation [oral contraceptives], and progestin, as in some progestin-only contraceptives, have been shown to lower the brain serotonin levels by increasing the concentration of a brain enzyme that reduces serotonin.[130]

The converse is also true: anti-depressant drugs that target se-rotonin uptake, such as Prozac and Zoloft, are known to have an adverse effect on libido.[131] They can also cause erectile dys-function and anorgasmia; one study in the 1990s found that 45% of female patients on SSRIs (selective serotonin reuptake inhibi-tors) experienced drug-induced sexual dysfunction and some studies suggest even higher rates.[132] This occurs because drugs that stimulate serotonin uptake can interfere with the uptake of hormones involved in sexual desire and reproduction, like do-pamine.[133] In other words, the issue cuts both ways: drugs that are or target hormones involved in sex can cause depression; drugs that treat depression can affect the hormones involved in sex.

We have known for fifty years that the Pill can cause mood disorders in women. We know that drugs that treat mood dis-orders can affect hormones involved with libido, and scientists know at least one mechanism by which this occurs. And a recent study was stopped because hormonal contraceptive caused mood disorders in male subjects. A reasonable person might therefore ask: what was left to be established? Or as my daughter put it, why was the finding that the Pill causes depression in women viewed as *news*?

Let us return to the Denmark study. It did not find that previ-ous studies of oral contraceptives had shown that hormonal contraception did *not* cause mood changes. Rather, it concluded that "inconsistent research methods and lack of uniform assess-ments [made] it difficult to make strong conclusions about which . . . users are at risk for adverse mood effects."[134] In other words, it suggested that until now, we didn't know enough to draw a firm conclusion.

These researchers took the conventional approach of assum-ing no effect and requiring statistical proof at a specific significance level to say that an effect had been detected—and was therefore *known*. So did the various studies that preceded them. There's nothing particularly shocking about this; it is common statistical practice. But it says, in effect, that if evidence is not available that meets that standard, we must conclude that our results are inconclusive—or in lay terms, that we just don't know.

There are two problems with this approach. The first, which is a general one, is that a negative finding is often taken as indicating "no effect," when in fact it simply means that the researchers have not been able to detect the effect, at least not at a level that achieves statistical significance. (Many negative studies actually do see effects, but not ones that pass the bar of statistical significance at the 95% level.)[135] It is the classic conflation of absence of evidence with evidence of absence, and it can lead to false negative conclusions. Still, if enough good studies are done that consistently fail to find an effect (or one really large one with great statistical power), we might fairly conclude that the effect really isn't there.

But what if there is evidence from non-statistical sources, such as patient reports, that there may well be an effect? What if there is a theoretical reason (as there is here) to think that an effect is in fact *likely*? In that case, why are we assuming that there is none? Why are researchers playing dumb? If we know or have reason to suspect that something is a risk, it may be warranted to flip the null and use a default assumption of "effect" rather than "no effect," or to accept a lower level of statistical significance. (This has sometimes been done, as when the Environmental Protection Agency accepted some studies of the impacts of secondhand smoke at a confidence level of 90% rather than 95% on the grounds that the same chemicals that were known to cause cancer in primary smoke were also present in secondhand smoke.)[136] After decades of case reports, and with a mechanism to explain why it might be so, researchers should have accepted the null hypothesis that the Pill could cause depression and sought statistical evidence to disprove that hypothesis.

The second problem relates to how we think about causation. The classic argument that correlation is not causation is misleading. What we should say is that correlation is not *necessarily* causation. Many things are correlated that are not causally related. But if we have an observed correlation between two phenomena, and we are aware of a mechanism that explains how one of them can be caused by the other, and if that mechanism is known to be present, then the logical conclusion is the observed correlation *is* caused by the known mechanism. Under these conditions, correlation *is* causation. Or at least, it is likely to be.

A classic example is the correlation between shark attacks and ice cream sales. Statisticians love to use this as an example to prove how correlations can be misleading: both are related to warm weather, when people swim in the ocean and eat ice cream. Neither one causes the other. But what if we had independent evidence that the smell of ice cream attracted sharks? Then it might be the case that there was a causal relation. Now suppose that the correlation did not achieve statistical significance at the 95% level. Would we conclude that there was no relation between the ice cream and the attacks? Under currently prevailing norms, we would. And we would be wrong. We need to pay attention to mechanisms.[137]

Consider another example. When the United States reduced the speed limit on interstate highways to fifty-five miles per hour, traffic fatalities dropped dramatically. The motivation for the speed limit change was to save fuel, not lives, so one might initially suppose that this correlation was just coincidental. In fact, driving at lower speeds reduces the chance of an accident and the likelihood that any accident that occurs will be fatal. Because we understand this, we rightly conclude that lowering the speed limit caused a decrease in traffic fatalities.

Playing dumb makes sense when we have no reason to suspect that phenomena are linked, or have affirmative reason to suppose they are not. If we knew nothing about hormones and mental health we might have rightly said that we needed more evidence to conclude that the Pill might cause depression. But we know that hormones affect brain chemistry. This is one reason why manufacturers have worked to decrease estrogen levels in oral contraceptives.

Women have always known that we sometimes get moody and depressed right before our periods. Indeed, popular lore makes us unreliable—as scientists, as political leaders, as CEOs—*because* of this. Stereotypes typically draw the wrong conclusions from the evidence on which they are based, but that in and of itself is not a refutation of the evidence. Hormones affect our moods. This is true for men and women.

Doctors who have not warned their patients of this risk during the past thirty years have been ignoring evidence. Public health officials who have downplayed the risk by discounting evidence—in this case, reams of it collected over more than three decades—because it did not meet certain methodological preferences have done women a grave disservice. Had doctors and public health professionals paid more attention to "iffy" case reports instead of discounting them, they would not simply have come to a better conclusion, epistemologically. They would have done their jobs better—and served their patients well—by not discounting a real and troubling side effect of an otherwise desirable medication. Given that the Pill has been implicated in suicidal ideation, they might even have saved lives.

Example 5: Dental Floss

My final case involves a very grave public health issue: dental floss.

Many people have recently heard that flossing your teeth doesn't do you any good. In August 2016 there was a flurry of coverage saying so. The *New York Times* asked, "Feeling Guilty about Not Flossing? Maybe There Is No Need."[138] The *Los Angeles Times* reassured its readers that if they didn't floss, they needn't feel bad because it probably doesn't work anyway.[139] So did *Mother Jones*, which ran the headline, "Guilty No More: Flossing Doesn't Work."[140] *Newsweek* asked, "Has the Flossing Myth Been Shattered?"[141]

These various reports were based on an article by the Associated Press (AP) that claimed that there is "little proof that flossing works." The AP quoted National Institutes of Health dentist Tim Iafolla, acknowledging "that if the highest standards of science were applied in keeping with the flossing reviews of the past decade, 'then it would be appropriate to drop the floss guidelines.'"[142] The *Chicago Tribune* linked this latest reversal in scientific fortune to previous (alleged) reversals on salt and fat.[143] Evidently, we can add dental floss to the list of issues on which scientists have "got it wrong."

It was not just the alleged lack of evidence that caught reporters' attention; there was also a suggestion of incompetence or even malfeasance. The *New York Times* suggested that the federal government may have violated the law that stipulates that federal dietary guidelines must be based on scientific evidence. So too the AP: "The federal government has recommended flossing since 1979, first in a surgeon general's report and later in the Dietary Guidelines for Americans issued every five years. The guidelines must be based on scientific evidence, under the law."[144] *The Week* ran the story under the headline "Everything You Believed about Flossing Is a Lie."[145] The *Detroit News* referred to the defenders of floss as the "floss-industrial complex."[146] One website called it "The Great Dental Floss Scam."[147]

Many reports contained an element of schadenfreude: some journalists seemed practically gleeful that journalists had one-upped scientists.[148] WRVO, an NPR affiliate in Oswego, New York, ran the story under the headline "How a Journalist Debunked a Decades-Old Health Tip."[149] The report claimed that the story began when AP reporter Jeff Donn learned from his "son's orthodontist . . . that there was in fact no good evidence that dental floss helps prevent cavities and gum disease."[150] Poynter.org labeled the story "How a Reporter Took Down Flossing."[151] A website promoting collective consciousness and natural living ran the story under the headline "The Deceit of the Dental Health Industry," stating that "flossing has been shown to be almost useless in terms of its purported benefits" and suggesting that "most of your oral health is determined by your diet and nutrition."[152]

On the face of it, this certainly appears to be a case of scientists having "got it wrong." Dentists and public health officials, including those in positions of governmental authority, have been instructing us about something that now we are told is not the case. We have wasted time and money on something useless. And this bears directly on the issue of trust, because if scientists have been wrong for decades about dental floss—as well as perhaps fat and sugar—then what else have they been wrong about? Will they tell us next that it is all right to smoke? Or that climate change *is* a hoax? Scientists might be tempted to respond that the fracas over flossing is just the messy work of science correcting itself, as a major scientific study revealed the weaknesses in previous work. But that is *not* what transpired. In fact, this is not a case of scientists getting it wrong at all. It's a case of journalists getting it wrong, and scientists getting blamed.

The "scientific" finding was not a scientific finding at all, but the result of an investigation by one reporter for the AP.[153] The source of the media story was the media itself. According to their own reporting, which they filed under the rubric "The Big Story," the AP "looked at the most rigorous research conducted over the past decade, focusing on 25 studies that generally compared the use of a toothbrush with the combination of toothbrushes and floss. The findings? The evidence for flossing is 'weak, very unreliable,' of 'very low' quality, and carries 'a moderate to large potential for bias.' 'The majority of available studies fail to demonstrate that flossing is generally effective in plaque removal,' said one review conducted last year. Another academic review, completed in 2015, cites 'inconsistent/weak evidence' for flossing and a 'lack of efficacy.' "[154]

The *New York Times*, seemingly staying close to the facts, informed readers that "A review of 12 randomized controlled trials published in The Cochrane Database of Systematic Reviews in 2011 found only 'very unreliable' evidence that flossing might reduce plaque after one and three months. Researchers could not find any studies on the effectiveness of flossing combined with brushing for cavity prevention." (We will address the distinction between gum health and cavity prevention in a moment.) But, as the *Times* rightly noted, that study was done in 2011, so how and why did this become a story in 2016?

According to the AP, their investigation of the matter was triggered by a decision by the US government to drop flossing from federal dietary guidelines (and not by Jeff Donn's conversation with his son's orthodontist, raising further questions about the whole story). This led them to ask the question: "What were those guidelines based on in the first place?"[155] While it was later revealed that the guideline change was the result of a decision to focus the dietary guidelines on diet—i.e., food—rather than other health practices, the cat was out of the bag.[156] The "finding" that flossing does not work was all over the news. As for Donn, he was quoted in a subsequent interview: "I think the best science indicates that [by flossing] I'm not doing anything beneficial for my health."[157]

Let us step back from the media coverage to ask: what scientific evidence exists to support or refute the claim that dental floss is of value? Donn is neither a scientist nor a dentist, and in fact his claim is not correct. The available science does *not* indicate that by flossing we are "not doing anything beneficial" for our health.

The most well-known and respected source of information on the state of the art in biomedicine is the Cochrane group, a nonprofit collaboration that bills itself as "representing an international gold standard for high quality, trusted information." The collaboration claims thirty-seven thousand participants from more than 130 countries who "work together to produce credible, accessible health information that is free from commercial sponsorship and other conflicts of interest."[158] As the *New York Times* correctly reported, in 2011, the collaboration issued a report from its oral health group reviewing existing clinical trials examining the benefits of regular use of dental floss.[159]

The report was based on a review of twelve trials, with 582 subjects in flossing-plus-toothbrushing groups and 501 participants in toothbrushing-alone groups. The report summary reads as follows:

> There is some evidence from twelve studies that flossing in addition to tooth-brushing reduces gingivitis compared to

tooth-brushing alone. There is weak, very unreliable evidence from 10 studies that flossing plus tooth-brushing may be associated with a small reduction in plaque at 1 and 3 months. No studies reported the effectiveness of flossing plus tooth-brushing for preventing dental caries [tooth decay].[160]

That part of the summary was reported, wholly or in part, in many of the media reports. But the report also said:

Flossing plus tooth-brushing showed a statistically significant benefit compared to tooth-brushing in reducing gingivitis at the three time points studied [although the effect size was small].[161] The 1-month estimate translates to a 0.13 point reduction on a 0 to 3 point scale for . . . gingivitis . . . and the 3 and 6 month results translate to 0.20 and 0.09 reductions on the same scale.[162]

This additional information refutes much of the media presentation. The crux of the news coverage was that many existing studies are weak, involving small numbers of people or very short periods of time. That is true. But it is not the same as demonstrating that flossing has no benefit. On the contrary, if the Cochrane review is correct, these studies indicate that, over the time period of the study, small but statistically significant reduction of gingivitis was observed in patients who flossed along with brushing.

The Cochrane review also considered evidence that flossing may help reduce plaque, which is associated with cavities as well as other matters. On this, they concluded that

Overall there is weak, very unreliable evidence which suggests that flossing plus tooth-brushing may be associated with a small reduction in plaque at 1 or 3 months. None of the included trials reported data for the outcomes of caries, calculus, clinical attachment loss, or quality of life.

Here we can identify one source of difficulty and potential misunderstanding: a number of different questions are being conflated, including whether flossing improves your life. Let us concentrate on the two main issues, as reported on both by the Cochranes and the news media: plaque and gingivitis. Plaque matters because it can lead to dental caries, and gingivitis matters because it is the first stage of periodontal disease, which can lead to tooth loss later in life. More than 70% of Americans over 65 have some form of periodontitis, which is *always* preceded by gingivitis.[163] If flossing reduces gingivitis, then it is likely that flossing reduces periodontal disease. Periodontal disease has been linked to serious illness, including increased risk of cancer and Alzheimer's disease.[164]

Dental floss defenders made this point. What the Cochranes concluded was not that flossing doesn't help, but that we don't have sufficient studies of high enough quality pursued over sufficiently long periods to demonstrate that it does help. The American Academy of Periodontology pointed out that "the current evidence fell short because researchers had not been able to include enough participants or 'examine gum health over a significant amount of time.' " Dr. Philippe Hujoel, a professor of oral health sciences at the University of Washington, Seattle, called it "very surprising" that "we don't have the . . . randomized clinical trials to show [flossing is] effective," given how widespread the belief is that flossing does help.[165]

But it is so surprising? Perhaps not. What we learned in 2016 was that we didn't have the long-term, randomized clinical trials that would be necessary to prove the benefits of dental floss according to prevailing medical standards. It's not that hard to understand why, in a world of cancer, heart disease, opioid abuse, and the continued use of tobacco products, such studies have not been done. It's not egregious that researchers have focused their attention on matters that appear to be more serious. What is egregious is that in the absence of evidence that meets the "gold" standard of the randomized clinical trial, people have concluded that there is no evidence at all. That is both false and illogical.[166]

Moreover, the gold standard of clinical trials is not just the *randomized* trial, but the *double-blind* randomized trial, and it is impossible to do a double-blind trial of dental floss. (This difficulty also plagues studies of nutrition, exercise, yoga, meditation, acupuncture, surgery, and any number of interventions of which the subject is necessarily aware.) Any study of floss usage will also require self-reporting, which, as we have seen, is disparaged. Moreover, if you believe that long-term flossing can prevent tooth loss in old age, it would be unethical to ask a control group to refrain from flossing for what would have to be the better part of their lives. The sort of study that would be required to convince those who subscribe to the "gold standard" is both impossible and arguably unethical to perform.[167]

Donn interpreted his findings to say that existing studies show no long-term benefits even when floss is used properly; once again we are observing the fallacy of equating absence of evidence with evidence of absence.[168] None of these studies was long enough to demonstrate long-term benefits. Dunn was also quoted as saying that there was "no good evidence." Whether this is correct depends on your definition of "good," but clearly there *is* evidence that flossing may have benefit.

In the aftermath of the negative media coverage, dentists who support flossing appealed to clinical experience. Several articles quoted dentists, professors of dentistry, and deans of dental schools affirming that clinical practice reveals that those who floss have healthier teeth and gums than those who don't. Some dentists went so far as to suggest that they can tell who among their patients is lying about their flossing habits simply by observing the conditions of their gums. (This reminds us of another reason why a good clinical trial would be hard to do: people lie about flossing. One study concluded that one in four Americans who claimed to floss regularly was fibbing.)[169] And then there is the experience of patients—which is to say, all of us. Many of us have noticed that when we floss regularly our gums don't bleed, and bleeding gums can be a sign of early periodontal disease. The dean of the dental school at the University of Detroit, Mercy, used this clinical and patient experience to suggest why high-quality trials had never been done: "They don't do research on things that are common knowledge."[170]

How can we reconcile the experience of dentists and patients with the lack of high quality, long-term epidemiological evidence? We could dismiss these observations as correlation but not causation, but we could also view the experience of dentists and patients as a form of observation that confirms the hypothesis that flossing helps prevent gum disease. In other words, as in the case of the Pill, we can accept the experience of patients and clinicians as evidence, even if the explanation for that evidence is not fully clear. Put another way, we can reject the rejection of this evidence as "merely" anecdotal, and insist that these are *case reports*, and n is far greater than 1. Moreover, as with the Pill, we can consider mechanism.[171] There is in fact good reason to think that dental floss is likely to be beneficial—that these correlations are in fact causally related—because it removes plaque and tartar that can contribute to gum disease, which, over time, can lead to tooth loss. Just as there is a known mechanism that links estrogen to serotonin and mood control, there is a known mechanism by which flossing is expected to prevent tooth loss.

This was explained by Dr. Sebastian G. Ciancio, the chairman of the department of periodontology at the University at Buffalo: "Gum inflammation progresses to periodontitis, which is bone loss, so the logic is if we can reduce gingivitis, we'll reduce the progression to bone loss." But severe periodontal disease may take five to twenty years to develop, so this effect cannot be demonstrated in a clinical trial that lasts only weeks or months. Dr. Wayne Aldredge, president of the American Academy of Periodontology put it this way: "It's a very insidious, slow, bone-melting disease. . . . You don't know if you'll develop periodontal disease, and you can find out too late."[172] In short, the "gold standard" of the randomized clinical trial is unable to reveal the benefits that periodontists predict. The clinical trials that have been undertaken were not the right tools for addressing that question.

The term "gold standard" should remind us that there are silver and bronze standards, too—or at least there should be. As Nancy Cartwright and Jeremy Hardie have argued, the ideal of a uniform gold standard is misguided: No one would use gold for household pipes; it is too expensive. Nor would we use gold for cooking knives: it is too soft.[173] The best tool depends on the job, and that applies to intellectual jobs as well as industrial and household ones.

What would be the right tool to investigate dental floss? One might be a different sort of clinical trial. The American Dental Association notes that disappointing results might be the result of poor flossing, which, they noted, is a "technique sensitive intervention."[174] The *New York Times* concluded: "So maybe perfect flossing is effective. But scientists would be hard put to find anyone to test that theory."[175] With due respect, that is an ill-informed remark, because scientists *have* tested that theory. The clinical trials reviewed by the Cochranes did not examine the impact of flossing technique, but a review of six trials in which professionals flossed the teeth of children on school days for almost two years, saw a *40% reduction in the risk of cavities.*[176]

That is a huge effect. So consider this alternative headline: "A New Job Opportunity: Science Shows the Need for Professional Flossers." Imagine the social change that might have ensued and the employment opportunities created. On our way to work, instead of stopping at Peets or Starbucks for a quick latte or a Drybar for a blow-out, we could stop at a flossing bar for a five-minute professional floss.

What Does It Take to Produce Reliable Knowledge?

There are many ways in which scientists can fail to live up to their own standards, as well as ways in which the standards they set can be unhelpful, incomplete, inadequate, or inappropriate to a particular situation. Still, I believe there are some themes that we may glean from these diverse cases. They are: (1) consensus, (2) method, (3) evidence, (4) values, and (5) humility.

Consensus

In chapter 1, we saw that historians, philosophers, and sociologists have come to focus on scientific consensus because there is no independent measure of what scientific knowledge is. We cannot identify science by any unique method. We can only identify claims as being scientific based on their provenance, that is to say, based on the way they were established and by whom. Scientific facts are claims about which scientists have come to agreement.

Some skeptics have used this argument to try to discredit contemporary science, claiming that there was a consensus supporting eugenics or rejecting continental drift.[177] This, they argue, proves that scientific consensus is an insufficient basis to command our trust, a faulty foundation for decision-making. But these claims are misplaced: scientists did *not* have a consensus about eugenics or continental drift. Social scientists, socialist geneticists, and some mainstream geneticists critiqued eugenics; the rejection of continental drift was a distinctly American affair. (Europeans for the most part withheld judgment, which is a different thing.) Nor was there a consensus over the Limited Energy Theory, the Pill, or dental floss. Gynecologists liked the Pill for its efficacy; psychiatrists were concerned about its psychological health side effects. Short-term epidemiological studies fail to find strong evidence for beneficial effects of flossing, but nearly all clinicians observe benefits. And leading women physicians pointed out the obvious flaws in the Limited Energy Theory.

A key finding from historical inquiry into these episodes, then, is that in all of these cases there was significant, important, and *empirically informed dissent within the scientific community*. When we see disputes within scientific communities across geographic, disciplinary, or other gaps, this should command our attention. Debates may arise between different types of scientific experts examining a common topic—psychiatrists and gynecologists— or between different types of people—male doctors and female ones—or between scientists in the same field bringing different background assumptions and values. These debates occur because different groups of scientists are emphasizing different bodies of evidence, highlighting different aspects of those bodies of evidence, or bringing different values and background assumptions into the interpretation of evidence.

Scientific consensus is hard to come by. This is an underappreciated fact. Therefore, in any debate, it is crucially important that we evaluate whether an expert consensus prevails or not. In 2004 I wrote a paper asking: Is there a scientific consensus on anthropogenic climate change? I had discovered that no one had analyzed the scientific literature with this question in mind and it seemed to me that any discussion of a mooted question should begin with an analysis of this sort.

In a recent issue of the *Hedgehog Review*, the editors wrote that "when we hear conflicting scientific pronouncements being issued on almost any subject (climate change, diet, vaccination) . . . it is not hard to see why science, and particularly scientific authority, has become the target of heated contestation and debate."[178] This claim is wrong on two counts. First, it has cause and effect backward. These issues are contested because various groups—the tobacco and fossil fuels industries, advocates of deregulation, parents of autistic children who feel inadequately supported, some evangelical Christians—are unhappy with scientific authority. Some of them *want* science to be devalued.

Because science has challenged their interests or beliefs, they challenge science. Contestation is the outcome of a conflict about authority. Second, these are not conflicting *scientific* pronouncements. On most of the scientific issues that are highly contested in American culture—evolution, vaccine safety, climate change—there is a scientific consensus. What is lacking is cultural acceptance by parties who have found a way to challenge the science. This is the source of the contestation, not conflicting positions within the scientific community. Political and cultural debate is by no means illegitimate, but political debate masquerading as science is dishonest. It has led to the sort of confusion displayed by the editors of *Hedgehog Review* and many others.

Consensus analysis of peer-reviewed literature (as I have done) is a means to determine whether scientists agree. If they do, then we can take the next step to identify who is contesting their findings, and why. In our book, *Merchants of Doubt*, Erik Conway and I were able to show that climate science was being contested by the fossil fuel industry, whose economic interests were threatened, and by Libertarian think tanks and conservative scientists whose political beliefs were challenged. Rather than admit this, they challenged the science as a means to protect their economic interests and political commitments.

If there is informed dissent within the scientific community, more (scientific) research may well be needed. If, however, the dissent is emanating from outside the relevant expert scientific community, then we have a different issue at stake. In the latter case, more scientific research is unlikely to settle the matter, because non-scientific objections are not driven by scientific considerations and therefore will not be resolved by more scientific information.

This is not to say that non-scientific objections are invalid, but only that they should not be confused with "scientific pronouncements." There can be important moral objections to social programs based on science, even if the underlying science is legitimate. And, as the contraceptive pill case illustrates, relevant information can emerge from outside specialist communities. My intent in presenting the Pill case was not to say that patients were necessarily correct, but rather that they had relevant information that should not have been disparaged simply because it came in the form of self-reporting.

How do we judge if non-experts have relevant, useful, and accurate information? This is not an easy question to answer. We have clear markers of scientific training and expertise: higher education, membership in scientific and learned societies, records of publication and research grants, H-indices, awards and prizes, and the like. Scientists know who their scientific colleagues are and what their track records look like. Scientists (for the most part) know which journals have rigorous peer review and which do not.

Judging information from outside the expert world, however, is a different and trickier matter.

Scholars have identified several categories worthy of attention. One is other professionals who have relevant information. This could include nurses and midwives, for example, who have direct contact with patients and may differ from physicians on questions such as pain management.[179] A second category is people who may not have professional training, but whose daily experiences may lead them to relevant knowledge and understandings, such as farmers and fishermen.[180] We might say that these people have daily "on the ground" experience, and therefore may see things that scientific experts, for whatever reason, have missed. (Earth scientists call this "ground truth," in this case referring to what geologists on the ground see and therefore know about, as compared with evidence, for example, from satellite remote sensing.) As Brian Wynne has stressed, the nonexpert world is not "epistemically vacuous."[181]

A third category is what Marjorie Garber has called "amateur professionals."[182] These are people—perhaps independent scholars or scholars from other fields—who have educated themselves on a particular subject. Developing expertise outside of conventional avenues of credentialism is certainly possible (although if a scholar from one field moves into another, they can establish credentials by publishing). A fourth category is citizen scientists: people who earn their living in other ways, but participate in science out of love or interest. In some domains—astronomy, entomology, ornithology, and the search for extraterrestrial life—citizen scientists have played significant roles in observing things that professionals do not have the time, money, or human resources to track.

People in all these categories may have knowledge relevant to a particular scientific question. They have a known relation to their object of study and basis for claiming a role in scientific conversations bearing on that study. Where their experience and expertise overlap with scientific expertise, we should pay attention and not automatically discount what they have to say, nor assume that their claims are necessarily in conflict with those of scientific experts.[183] Often expert and lay perspectives can be reconciled or seen as complementary. But, again to draw on Wynne, we should not misunderstand a claim for recognition of these knowledge categories as a claim for their intellectual superiority or equivalence.[184] Just because someone is close to an issue does not mean he or she understands it; conventional notions of objectivity assume distance for just this reason. Parents of autistic children will have detailed knowledge of their children's conditions, but this does not mean that they are in a position to judge what caused it.[185]

Respecting professional diversity and lay expertise is also a different matter from heeding "dissent" from people with no credible claim to expertise—celebrities, K-Street lobbyists, or the op-ed writers of the *Wall Street Journal* or the *New York Times*. When people without relevant expertise criticize science, we should consider the possibility that something fishy may be going on. If people are attacking science, there is something at stake, but it is not necessarily something scientific. Indeed, it is probably not.

An abundant literature now documents how various parties have tried to create the impression of scientific uncertainty and debate as a means to block public policy that conflicts with their

political, economic, and ideological interests.[186] But these are not
the only reasons that people attack science, insist there is no consensus, or promote alternative theories. People attack science
to get attention, to sell alternative therapies, or because they are
frustrated that science doesn't have an answer to a problem that
affects them.[187] But it is a relatively simple matter to distinguish
between scientific debate and other stuff: Scientific debate takes
place within the halls of science and on the pages of academic
journals; other stuff takes place in other places. Political debate
takes place on the op-ed pages of newspapers. Grievances can
be aired anywhere. Sadness, isolation, and frustration make
people lash out. But if, like the editors of *Hedgehog Review*, we
mischaracterize political debate, industry shilling, or social dis-
affection as scientific controversy, then our attempts to remedy
the situation will almost certainly fail.

Method

In the episodes we have been discussing, problems arose because
scientists discounted evidence that failed to meet their methodological preferences. In the early twentieth century, geologists
rejected continental drift, because it did not fit their inductive
methodological standards. Charles Davenport was attracted to
eugenics in part because he wanted to make biology more rigorous by making it more quantitative. In the cases of dental flossing and the Pill, scientists discounted clinical evidence because
of a lack of robust epidemiological data. This last point is particularly important, because in the contemporary world, we have
come to rely on statistical analysis to a degree that has led
many people to ignore important evidence, including the evidence of everyday experience that hormones affect our moods

and flossing makes our gums less bloody. This doesn't mean that everyday experience is superior to statistics; it is not. Good statistical studies are an essential part of modern science. It just means that statistics, like any tool, don't work well in all cases and conditions and like any tool can be used well or badly (Krosnick, this volume).

A focus on one method above all others is a kind of fetish. These cases suggest that some of the historical examples of "science gone awry" arose from what I designate *methodological fetishism*. These are situations where investigators privileged a particular method and ignored or discounted evidence obtained by other methods, which, if heeded, could have changed their minds.

Experience and observation come in many forms. A good deal of evidence is imperfect, but that is no reason to ignore it. It is foolish to discount evidence that comes in messy forms simply because they are messy, particularly when the preferred methodological standard is difficult to meet or unsuitable to the question at hand. Randomized double-blind trials are powerful when they can be done, but when they cannot we should not throw our hands up and suggest we know nothing. There is no way to know how a drug makes people feel without asking about their feelings. There is no way to do a double-blind trial of flossing or nutrition. Imperfect information is still information.

When we have independent information about causes and mechanisms—such as knowing that flossing reduces gingivitis, that hormonal contraception can affect serotonin receptors (and vice versa: that anti-depressants that target serotonin uptake can affect hormones), or that greenhouse gases alter the radiative balance of the planet—this information is crucial to helping us evaluate claims when our statistical information is noisy, inadequate, or incomplete. Mechanisms matter. When we know something about relevant mechanisms, there is no reason to play dumb.[188]

Evidence

It seems obvious to say, but scientific theories should be based on evidence. However, in two of the cases here, we saw scientists making affirmative claims on the basis of little or scant evidence. Dr. Edward Clarke built an ambitious and socially consequential theory about female capacity on the basis of seven patients. Critics at the time noticed not only that his data base was scant, but also that it was biased: his patients were all young women who had come to him suffering anxiety, backache, headache, and anemia, and who he described as pursuing educational or professional goals in a "man's way."[189] (This included an actress and a bookkeeper; only one was actually a student in a woman's college.)

In hindsight it is more than obvious that the symptoms he described—headache, backache, anxiety—could have any one of a number of causes. They are also afflictions that often occur in men, yet Clarke offered no evidence that these ills were more common in women, or more common among women who were educated than in those who were not. He presented his theory in the framework of hypothetico-deductivism, yet he failed to pursue the required next step: to determine if his deduction were true. Most conspicuously, he provided no evidence that these women's reproductive systems were weakened or that their fertility had been decreased. When women physicians and educators pointed out these flaws, Clarke ignored them. His theory was elegant, but could only be sustained by ignoring evidence available to him at that time.

Values

The role of values in science is a much-mooted issue, and the stories told here show how easily prevailing social prejudices may be instantiated into scientific theory. Scientists have not always been on the side of the angels. Anyone who values science must acknowledge this.

The traditional impulse of scientists has been to say that in cases such as eugenics, science was "distorted" by values. But historians of science, particularly but not only feminists, have noted the ways in which values are broadly infused into scientific life and not always in adverse ways. It is true that racial and ethnic prejudice infused eugenic thinking, and the sexism in Edward Clarke's work is not difficult to discern. But values also played a role in the critiques of those theories. Socialist values were crucial to some geneticists' critique of eugenic thinking; feminist values informed Mary Putnam Jacobi's identification of the theoretical and empirical inadequacies of the Limited Energy Theory. Barbara Seaman was a journalist, not a scientist, but her feminist values motivated her to follow up on the "anecdotes" she had heard, to seek out the doctors who could confirm the substance in these stories, and to highlight information that some doctors were discounting.

This, it seems to me, is the most important argument for diversity in science, and for diversity in intellectual life in general. A homogenous community will be hard-pressed to realize which of its assumptions are warranted by evidence and which are not. After all, just as it is hard to hear your own accent, it is hard to identify prejudices that you share. A community with diverse values is more likely to identify and challenge prejudicial beliefs embedded in, or masquerading as, scientific theory.

Critics of efforts to make science more diverse sometimes insist that the only relevant standard in science is "excellence."[190] Science, they insist, is a meritocracy in which demographic considerations are misplaced. These critics seem to think that calls for diversity are *merely* political; that there is no intellectual value in building diverse communities. The stories told here refute that idea. They suggest that diversity can result in a more rigorous intellectual outcome by fostering critical interrogations that reveal embedded social prejudice.

Admittedly, this claim cannot be proved, because in science we have no independent metric to judge epistemic success. We cannot stand apart from our truth claims and independently determine if they are true; nor can we compare the "truth-production" of more and less diverse communities. But in a domain where there are metrics of success—namely, business—rigorous studies have demonstrated that diverse teams yield better outcomes, in terms both of qualitative values such as creativity and quantitative outcomes such as sales. If we know that diversity is beneficial in the commercial workplace, why would we not presume that it would be beneficial in the intellectual workplace as well? Moreover, we saw in chapter 1 that there is an epistemological basis for presuming that diversity does benefit science. The examples presented in this chapter support that claim. Thus we may conclude that scientific communities that that are "politically correct"—in the sense of taking seriously the value of diversity—are more likely to yield work that is scientifically correct.

Considering the role of values also helps explain what we could call the *misapplication of theory* and the *asymmetry of application*. In hindsight, there is an obvious theoretical flaw in Clarke's work: while presented as an application of thermodynamics, it was actually a misapplication of the theory because conservation of energy applies to closed systems. The human body is not a closed system: it is sustained and supported through nutrition. Life is possible *because* organisms are not closed systems, so Clarke's use of thermodynamics was logically fallacious. It was also asymmetrical, because for some odd reason it only applied to women. Admittedly, Clarke had an explanation for this: he suggested that the female contribution to reproduction was uniquely demanding, and he did allow the possibility that overexertion could be harmful to both boys and men as well. Yet, while stressing the claim that if a woman was educated, her uterus would shrink, he evidently never paused to ask: if men were educated, what part of their anatomy would shrink?

Eugenicists likewise applied their theories asymmetrically. As Muller and Haldane stressed, the target of their attention was the working class. There were drunkards, gamblers, and lay-abouts among the wealthy, yet few eugenicists advocated sterilization of underperforming rich white men.

Humility

If the history of science teaches anything, it is humility. Smart, hard-working, and well-intentioned scientists in the past have drawn conclusions that we now view as incorrect. They have allowed crude social prejudice to inform their scientific thinking. They have ignored or neglected evidence that was readily available. They have become fetishists about method. And they have successfully persuaded their colleagues to take positions that in hindsight we see as incorrect, immoral, or both.

Many of the scientists in these stories were driven by a genuine desire to do good: to promote an effective means of birth control, for example, or protect women from something they honestly believed would harm them. But their failings are a reminder that anyone engaged in scientific work should strive to develop a healthy sense of self-skepticism. Edward Clarke was a supremely confident man. So was Charles Davenport. So were many of the early advocates of the contraceptive pill. Wegener's critics accused him of "auto-intoxication," and I daresay we have all encountered scientists who are overly enamored of themselves. It seems to me that individual scientists, if they care about truth, should be mindful of this problem and not ride roughshod over their colleagues.

If the social view of science is correct, however, then it may not matter too much if a particular individual is auto-intoxicated. Inevitably there will be arrogant individuals in science, but so long as the community is diverse and alternative views are available, and so long as the community as a whole finds the means for all its members to be heard, things are likely to go well. Nonetheless, collectively scientists should still bear in mind that—whatever conclusions they come to and however they come to them—even with the best practices and the best of intentions, there is always the possibility of being wrong, and sometimes seriously so.

Conclusion:
Science as a Form of Pascal's Wager

In evaluating a scientific claim that has social, political, or personal consequences there is one more question that needs to be considered: What are the stakes of being wrong in either direction? What is the risk of accepting a claim that turns out to be false versus the risk of rejecting a claim that turns out to be true?

Knowing there is a risk of depression, if a healthy woman decides to take the Pill she can quickly stop taking it if the risk materializes. Pill-induced depression generally clears up quickly, so for many women the risk is modest and worth taking. Similarly, dental floss is cheap and only takes a few minutes a day to use. If it turns out to have little benefit, little has been lost. But some issues are not so easily resolved.

Consider anthropogenic climate change. Despite fifty years of sustained scientific work, communicated in tens of thousands of peer-reviewed scientific papers and many hundreds of governmental and nongovernmental reports, many people in the United States are still skeptical of the reality of climate change and the human role in it. The president has doubted it, as have members of Congress, business leaders, and the editorial page of the *Wall Street Journal*. Rejecting centuries of well-established physical theory and reams of empirical evidence regarding matters such as sea level rise and the intensification of extreme weather events, others have suggested that while anthropogenic climate change might be a real thing, it is inconsequential and might even be beneficial.[191]

As a historian of science, mindful of the Limited Energy Theory and eugenics and the history of hormonal contraception—mindful of the difficulties of evaluating dental floss—and above all mindful of the political ideals that geologists brought to bear in evaluating continental drift—I have never *assumed* that trust in science is always or even usually warranted. I have always felt that it is fair to ask: What is the basis for any scientific claim? *Should* we trust scientists?

We cannot eliminate the role of trust in science, but scientists should not expect us to accept their claims solely on trust. Scientists must be prepared to explain the basis of their claims and be open to the possibility that they might be wrongly dismissing or discounting evidence. If someone—be it a fellow scientist, an amateur professional, a journalist, or an informed citizen—has a credible case that evidence is being discounted or weighed asymmetrically, this should concern us. Scientists need to remain open to the possibility that they have made a mistake or missed something significant.[192] The key point is that the basis for our trust is not in scientists—as wise or upright individuals—but in science as a social process that rigorously vets claims.

This does not mean that scientists must spend time and energy continuing to prove and reprove conclusions that have already been established beyond a reasonable doubt, nor refuting claims that have been refuted. As Thomas Kuhn argued more than half a century ago, to the extent that science can be said to progress, it is because scientists have mechanisms by which they reach agreement and *then move on*. Perhaps the most salient aspect of the continental drift debate is that it *was* reopened, which occurred when a new generation of scientists developed new lines of pertinent evidence.[193]

We can reframe this problem in terms of Pascal's Wager. No matter how well-established scientific knowledge is—no matter how strong the expert consensus—there will always be residual uncertainty. For this reason, if our scientific knowledge is being challenged (for whatever reason), we might take a lead from Pascal and ask: What are the relative risks of ignoring scientific claims that turn out to be true versus acting on claims that turn out to be false?[194] The risks of not flossing are real, but not inordinate. The risks of not acting on the scientific evidence of climate change are inordinate.[195]

Admittedly, the advocates of eugenic social policies considered the risks of not implementing eugenic social policies to be extremely high. That, of course, was their interpretation of the scientific evidence. But as we have seen, there was no consensus on that *evidence*. So we are back to the importance of consensus. If we can demonstrate that there is no consensus among relevant experts, then it becomes clear that we have a weak basis for public policy. This is the reason why the tobacco industry tried for so long to claim that the science regarding the harms of tobacco was unsettled: if it really had been, then they might have been right to insist that tobacco control was premature.[196] Similarly, if there were no scientific consensus about anthropogenic climate change, then the fossil fuel industry and Libertarian think tanks might be right to ask for more research. This is why consensus studies are relevant and important: Knowing there is a consensus does not tell us what to do about a problem like climate change, but it does tell us that we almost certainly have a problem.[197]

If we can establish that there is a consensus of relevant experts, then what? Can we be confident in accepting their conclusions and using them to make decisions? My answer is a qualified yes. Yes, *if* the community is working as it ideally should. That is a substantial qualification. As Brian Wynne has put it, if we are to respect and trust science, then "it becomes evident why the quality of its institutional forms—of organization, control and social relations—is not just an optional embellishment of science in public life, but an essential component of critical social and cultural evaluation."[198]

The history of science shows that there is no guarantee that the ideals of an open, diverse community, participating in transformative interrogation, will be achieved. Often it will not be (although the consequences of failing to meet this ideal may not always be profound or even significant). Historian Laura Stark notes that the National Bioethics Advisory Commission recommends that one-quarter of the members of the boards that review human subjects research should not be affiliated with the institution at which the research is being done, but this goal is rarely achieved.[199]

How do we determine if a scientific community is sufficiently diverse, self-critical, and open to alternatives, particularly in the early stages of investigations when it is important not to close off avenues prematurely? How do we evaluate the quality of its institutional forms? We must examine each case on an individual basis. Many scientists were wrong about continental drift, but that does not mean that a different group of scientists are wrong today about climate change. They may be or they may not be. We cannot assert either position a priori.

If we can establish that there is a consensus among the community of qualified experts, then we may also want to ask

- Do the individuals in the community bring to bear different perspectives? Do they represent a range of perspectives in terms of ideas, theoretical commitments, methodological preferences, and personal values?
- Have different methods been applied and diverse lines of evidence considered?

- Has there been ample opportunity for dissenting views to be heard, considered, and weighed?
- Is the community open to new information and able to be self-critical?
- Is the community demographically diverse: in terms of age, gender, race, ethnicity, sexuality, country of origin, and the like?

This latter point is needs further explication. Scientific training is intended to eliminate personal bias, but all the available evidence suggests that it does not and probably cannot. Diversity is a means to correct for the inevitability of personal bias. But what is the argument for *demographic* diversity? Isn't the point really the need for *perspectival* diversity?

The best answer to this question is that demographic diversity is a proxy for perspectival diversity, or, better, a means to that end. A group of white, middle-aged, heterosexual men may have diverse views on many issues, but they may also have blind spots, for example, with respect to gender or sexuality. Adding women or queer individuals to the group can be a way of introducing perspectives that would otherwise be missed.

This is the essential point of standpoint epistemology, raised particularly by the philosopher Sandra Harding (chapter 1). Our perspectives depend to a great extent on our life experience, so a community of all men—or all women for that matter—is likely to have a narrower range of experience and therefore a narrower range of perspectives than a mixed one. Evidence from the commercial world supports this point. Studies of gender diversity in the workplace show that adding women in leadership positions increases company profitability—but only up to a point. That point is about 60%. If a company's leadership becomes all or nearly all female, then the "diversity bonus" begins to decline, as indeed, if the argument here is correct, it should.[200]

It may not always be easy to answer the questions posed above in the affirmative, but it is often obvious if the answer to any of them is negative. Moreover, we may (and likely will!) identify individuals in the community who are arrogant, closed-minded, and self-important, but on the social view of epistemology the behavior of any particular individual is not what matters. What matters is that the group as a whole includes enough diversity and maintains sufficient channels for open discussion that new evidence and new ideas have a fair chance of a fair hearing.

The philosopher Heather Douglas has argued that when the consequences of our scientific conclusions are non-epistemic—i.e., when they are moral, ethical, political, or economic—it is almost inevitable that our values will creep into our judgments of evidence.[201] (Liberals, for example, may have been quicker to accept the scientific evidence of climate change, because they were more comfortable with its implied consequences for government intervention in the marketplace.) Therefore, the more socially sensitive the issue being examined, the more urgent it is that the community examining it be open and diverse.

But sometimes an issue that appears to be purely epistemic isn't, and scientists may claim they are evaluating an issue on solely epistemic grounds even when they are not.[202] This suggests that no matter what the topic, it behooves the scientific community to pay attention to diversity and openness in its ranks and to remain open to new ideas, particularly when they are supported by empirical evidence or novel theoretical concepts. It means, for example, that in considering dissenting views in grant proposals or peer-reviewed papers, it is probably better to err on the side of tolerance than critique. Many scientists consider it extremely important to be intellectually tough, if not actually rough, but sometimes toughness can have the unintended effect of shutting down colleagues, particularly those who are young, shy, or inexperienced. It *is* important to be tough, but it may be more important to be *open*.

In chapter 1, I argued that the advocates of the Extended Evolutionary Synthesis are receiving a thorough hearing, if not always a polite one. Something similar happened to Alfred Wegener. He was not a neglected genius: his papers were published in peer-reviewed journals and his work had a hearing—albeit not always a gracious one. The socialist opponents of eugenics likewise got their manifesto published in *Nature*.[203] None of these dissenters was "shut down" by the scientific hierarchies of their day.

Fast forward to the present: Many AIDS researchers lament Peter H. Duesberg, the University of California molecular biologist who does not accept that AIDS is caused by a virus. By his own account he has "challenged the virus-AIDS hypothesis in the pages of such journals as *Cancer Research, Lancet, Proceedings of the National Academy of Sciences, Science, Nature, Journal of AIDS, AIDs Forschung, Biomedicine and Pharmotherapeutics, New England Journal of Medicine and Research in Immunology*."[204] Whether he is right or wrong, tolerated or vilified, the fact is that he has had a hearing, and at the highest level of American and international science. His colleagues have not shut him down; they have published his work and considered his arguments. But they remained unconvinced.[205] There is a difference between being suppressed and losing a debate. Sometimes a "skeptic" is just a sore loser.

CODA

Values in Science

Some people worry that overconfidence in the findings of science or the views of scientists can lead to bad public policy.[1] I agree: overemphasizing technical considerations at the expense of social, moral, or economic ones can lead to bad decisions.[2] But this does not bear on the question of whether the science involved is right or wrong. If a scientific matter is settled and the scientific community that has settled on it is open and diverse, then it behooves us to accept that science and then decide what (if anything) to do about its implications.

This, at least, is what nearly every scientist I know would say. It is something that in the past I have said. It actualizes the classic fact/value distinction: the idea that we can identify facts and then (separately) decide what if anything to do about them based on our values. But as an empirical matter this strategy is no longer working (if it ever did), because most people do not separate science from its implications.[3] Many people reject climate science, for example, not because there is anything wrong with that science, qua science, but because it conflicts—or is seen as conflicting—with their values, their religious views, their political ideology, and/or their economic interests.[4] There are many reasons people may reject or be critical of scientific findings, but often it involves the perception that these findings contradict their values or threaten their way of life.

In the 1960s, many people on the political left criticized science because of its uses in warfare.[5] Today, many on the right critique it because of the way in which it exposes flaws in contemporary capitalism and the American way of life.[6] In a discussion of anthropogenic climate change prior to the 1992 Earth Summit in Rio de Janeiro, US president George H. W. Bush insisted that "the American way of life is not up for negotiation. Period."[7] The president signed the UN Framework Convention on Climate Change—the international treaty that emerged from the Rio meeting—and promised to act upon it. Yet, at the same time, he identified a clash between the implications of the findings of environmental science and the highly consumptive American lifestyle. At least some environmentalists were blaming that lifestyle for environmental ills and thus wanting to change it. This pattern persists and helps to explain why Republicans are so much more skeptical of climate science than Democrats.[8] Indeed, it is the only thing that explains why some conservatives insist that proposals to act on climate change are anti-democratic, anti-American, and/or anti-freedom.[9]

If we ask scientists, "Why do evangelical Christians reject evolutionary biology?" many would answer that it is because they make a literal reading of the Bible, insisting that God created the Earth and everything on it in six days. But as Catholic evolutionary biologist Kenneth Miller has pointed out, evangelical arguments against evolutionary theory rarely involve literal interpretations of the Bible. In fact, they rarely involve biblical exegesis at all.[10] Rather, they invoke the perceived moral (or amoral) implications of a theory that says humans arose by chance, the outcome of a nonpurposeful, random process. Former Pennsylvania senator and two-time presidential candidate Rick Santorum, for example, has explained that he rejects the concept of evolution by natural selection because it makes humans into "mistakes of nature," and in doing so obliterates the basis for morality.[11] Other anti-evolutionists argue that if evolution is true, then life has no meaning.

Scientists attempt to escape the sting of these extra-scientific considerations by retreating into value-neutrality, insisting that while our science may have political, social, economic, or moral implications, the science itself is value-free.[12] Therefore, values are not legitimate grounds on which to reject it. Gravity doesn't care if you are a Republican or Democrat. Acid rain falls on both organic farms and golf courses. Radiative transfer in the atmosphere functions today just as it did before the last election.

This argument is true but insufficient, because, whether or not they should, our audiences link science to its implications. Evangelical Christians reject evolutionary science because they believe it contradicts their religious beliefs. Evangelical free-marketers reject climate science because it exposes contradictions in their economic worldview. And because of these contradictions, they distrust the *scientists* responsible for them. This is hard to get around, particularly when we acknowledge that there is no singular scientific method that warrants the veracity of scientific conclusions, and that science is simply the consensus of relevant experts on a matter after due consideration

A view of scientific knowledge as the consensus of experts inevitably brings us to the question of who scientists are and on what basis they should be trusted. Scientists typically consider such questions to be ad hominem and therefore illegitimate. But if we take seriously the conclusion that science is a consensual social process, then it matters who scientists are.[13]

It is beyond the scope of this volume to consider what social scientists have discovered as to how trust is created and sustained, but one thing we know is that it is easier to establish trust among people with shared values than among those without them.[14] Yet values are precisely what scientists, by and large, decline to discuss. Like the question of who the scientist is, the question of what scientists believe in—other than science itself—has been considered to be off-limits. When their objectivity or integrity is questioned, scientists characteristically retreat, as sociologist Robert Merton noted decades ago, into the "exaltation of pure science," insisting that their only motivation is the pursuit of knowledge.[15] Whatever the implications of scientific findings, they insist that the enterprise itself is value-neutral.

Merton thought this made sense: he believed that public trust in science was directly linked to the perception of its independence from outside interests—what he called "extra-scientific considerations." This is the reason that scientists defend the "pure science ideal."[16]

> One sentiment which is assimilated by the scientist from the very outset of his training pertains to the purity of science. Science must not [in this view] suffer itself to become the handmaiden of theology or economy or state. The function of this sentiment is to preserve the autonomy of science. For if such extra-scientific criteria of the value of science as presumable consonance with religious doctrines or economic utility or political appropriateness are adopted, science becomes acceptable only insofar as it meets these criteria. In other words, as the "pure science sentiment" is eliminated, science becomes subject to the direct control of other institutional agencies and its place in society becomes increasingly uncertain. . . . The exaltation of pure science is thus seen to be a defense against the invasion of norms that limit directions of potential advance and threaten the stability and continuance of scientific research as a valued social activity.[17]

For Merton, value-neutrality as a regulative ideal was important not only in helping scientists to maintain objectivity in
their research practices, but also in helping to maintain the public perception of them as fair, objective, and dedicated to the
pursuit of truth (as opposed to the pursuit of power, money, status, or anything else). To the extent that scientists were seen to
be pursuing goals other than scientific ones, people could and
likely would distrust them. Merton therefore thought scientists
right to defend the pure science ideal, and he would likely have
been dismayed by the ways in which contemporary scientific
leaders promote economic utility as the primary rationale for
research, universities aggressively pursue private sector support
for "basic" research, and individual scientists embrace private
profit as a motive in creating biotech start-ups and other entre-
preneurial endeavors.

But Merton was a sociologist, not a historian, and his full-
throated defense of the pure science ideal sits in tension with the
historical reality that scientists have always had patrons with motivations of their own, and which only rarely involved the pursuit
of knowledge for its own sake. Seen this way, the idea of science
as a value-neutral activity is a myth.[18] Utility—economic or
otherwise—has long been a justification for the support of science, in terms of both finance and cultural approbation. Historian John Heilbron has demonstrated that the Catholic Church
supported astronomy in the middle ages because it needed astro-
nomical data to determine the date of Easter.[19] Merton himself
was known for the argument that modern science thrived—and
took its contemporary form—in seventeenth-century England
because its emphasis on utility resonated with dominant Puritan
values.[20] In my own work I have shown how the US Navy supported basic research in oceanography in the Cold War for its
value in anti-submarine warfare and subsurface surveillance.[21]

Biology has long been supported by governments and philanthropists because of its value in medicine and public health; geology for finding useful resources (and also for its power to deepen our appreciation of God); physics for its use in technology; climate science for its relation to weather forecasting.[22] To suggest that utility has no place in the motivation for science is to ignore centuries of history. And utility is inescapably linked to values: health, prosperity, social stability, and the like. To say that something is useful is to say that we value it, or that it preserves, protects, or fosters something that we value.

If science as an enterprise is not value-neutral, neither are scientists as individuals. No one can be truly value-neutral, so when scientists claim that they are, it comes across as false, for they are claiming the impossible. Unless we accept them as idiot savants or naïfs, we may come to see them as dishonest. Yet, honesty, openness, and transparency are said to be key values in scientific research. How can scientists be honest and at the same time deny that they have values? Scientists generate a contradiction at the root of their enterprise if, while insisting on its honesty, they mislead their audiences (even if unintentionally) about its character.

It may be objected that scientists are not claiming that they have no values, but only that they do not allow those values to influence their scientific work. That is a claim that is impossible to prove or disprove, but one that both social science research and common sense suggests is unlikely to be true. This leads us to a further point, one that somehow has escaped serious consideration but which may be at the heart of the distrust of science felt by many Americans. To say that science is value-neutral is more or less equivalent to saying that it has no values—at least none other than knowledge production—and this can elide into the implication that *scientists* have no values. Clearly this is not the case, but scientists' reluctance to discuss their values can give the impression that their values are problematic—and need to be hidden—or perhaps that they have no values at all. And would you trust a person who has no values?

In chapter 2, I posed the question: What are the relative risks of ignoring scientific claims that turn out to be true versus acting on claims that later turn out to have been incorrect? The answer to this question depends on values. As Erik Conway and I showed in *Merchants of Doubt*, fights about climate science have for the most part been fights about values. Some influential people in the 1980s and 1990s believed that the political risks of government intervention in the marketplace were so great as to outweigh the risks of climate change, and so they discounted, disparaged, or even denied the scientific evidence of the latter. As these positions became adopted by Libertarian think tanks, and then aligned with the Republican Party, it became normal for Republicans to engage in climate change denial, either actively or passively. Then climate change skepticism became normal for anyone suspicious of "big government," which meant many business people, older men, evangelical Christians, and people living in rural America. As the evidence of climate change was accumulating all around them, skeptics insisted that even if the climate were changing, it was not serious, or it was not "caused by us." Because if it were serious and we had caused it, then *we* would have to do something about it, and that something in some way or another would involve governance. Thus, climate change denial became normalized in American life, and with it the denial of evidence and ultimately of facts. This is a deeply troubling state of affairs, but the values that have underpinned climate denial cannot be summarily rejected as wrong or false.[23]

We can argue about the relative merits of government large and small, and the relative risks of under- versus over-regulation of markets, but any such argument will be (at least in part) value-driven. If we are to have such conversations honestly, then we must talk about our values. Different people may view the same risk differently, but this does not mean they are stupid or venal. The scientific evidence of anthropogenic climate change is clear—as is the evidence that vaccines do not cause autism and that flossing your teeth is beneficial—but our values lead many of us to resist accepting what the evidence shows.

To return to the question: Would you trust a person who has no values? The answer is obvious: you would not. Such as person would be a sociopath. Nor would you trust a person whose values you considered to be anathema to your own. But if you thought that person shared at least some of your values—even if perhaps not all of them—you might be willing to listen. And you might accept some of what you were hearing. Therefore, whether or not claims of value-neutrality are epistemologically defensible, it is *clear that they do not work in practice, because they do not work to permit communication and build bonds of trust.*

The dominant style in scientific writing is not only to hide the values of the authors, but to hide their humanity altogether. Not only are values unexpressed, emotions expunged, and adjectives eschewed, but the word "I" is implicitly forbidden, even in single-authored papers.[24] This is tied to the performance of objectivity: The ideal scientific paper is written not only as if the author had no values or feelings, but as if there were no human author at all.[25]

Scientists may feel that there is simply no way for them to forge bonds of trust with climate change deniers or people who think the world is six thousand years old. Perhaps that is true. I once despaired publicly of ˙reaching millenarians whose eschatology tells them the world is about to end, so why worry about climate change? But when I did so despair, the next day several correspondents offered me strategies to reach them based on Christian values and teachings.[26] People suggested that the way to reach people was through their values. And social science research supports this idea.[27]

Conclusion

In suppressing their values and insisting on the value-neutrality of science, scientists have gone down a wrong road.[28] They have made the mistake of thinking that people would trust them if they believed that science was value-free.

Merton certainly thought so, but he may have been wrong. It may be that the opposite is true. Here's why.

While the rejection of evolution or the denial of anthropogenic climate change has led analysts to focus on the value clashes between scientists and politically and socially conservative Christians, Libertarians, and Republicans, I believe that the values that motivate most scientists overlap with the values of most Americans, including many conservatives and religious believers. Recently, some scientists have begun to declare their values publicly, in part, I think, because they believe that these values *are* broadly shared and therefore offer a basis for building bonds of trust.[29] I think they are right.

Most scientists I know want to prevent disease and improve human health, strengthen the economy through innovation and discovery, and protect the natural beauty of America and the world. Former Republican congressman Bob Inglis has spoken eloquently of visiting the Great Barrier Reef with a marine biologist. As the two men, side by side, admired the stunning beauty of the life around the reef, Inglis had a realization: He saw "Creation," the scientist saw "biodiversity," but they were, in fact, looking at, caring about, and *cherishing* the same thing.

This is a good story, because most people appreciate nature in at least some way. National parks and forests are visited by large numbers of diverse Americans—hiking, fishing, camping, driving, photographing, wondering, complaining—but sharing some appreciation for what they are seeing and experiencing. Yet, there are genuine conflicts of value in our relationship to the natural world. The desire of some to ride snowmobiles through Yellowstone in winter conflicts with the desire of others for contemplative recreation. Nearly all Americans would affirm their belief in freedom, yet what we understand by that word and which particular freedoms we prioritize can vary greatly. As Isaiah Berlin memorably put it, freedom for wolves can mean death for lambs.[30] Agreeing on the word "freedom" only gets us so far.

Historian of religion Stephen Prothero has noted that while Jews, Catholics, and Protestants all affirm the Ten Commandments, there are surprisingly different versions of them.[31] Catholics, for example, abandoned the injunction against graven images onto which Jews and Protestants firmly held, and having thus lost a commandment—and finding themselves awkwardly left with only nine—they divided the last one into two, the ninth for coveting your neighbor's wife and the tenth for coveting everything else. Nonetheless, these three religions—which encompass about 70% of American adults—agree that we should not kill, steal, commit adultery, or bear false witness. They also agree that we should worship only one God, not take his [sic] name in vain, keep the Sabbath, and honor our fathers and mothers. Islam agrees with these, although it stresses charity more centrally than the other three do: *zakat*—the giving of alms—is one of its five pillars. Yet note how similar the word *zakat* is to the Hebrew word *tzedakah*—charitable giving—which is considered a moral obligation in Jewish life. Charity is a central Christian value as well. Observant Mormons tithe.

Even as we disagree about many political issues, our core values overlap to a great degree. To the extent that we can make those areas of agreement clear—and explain how they relate to scientific work—we might be able to overcome the feelings of skepticism and distrust that often prevail, particularly distrust that is rooted in the perception of a clash of values.

So let me be clear about my values.

I wish to prevent avoidable human suffering and to protect the beauty and diversity of life on Earth. I wish to preserve the joy of winter sports, the majesty of coral reefs, and the wonder of giant sequoia trees. I love thunderstorms, but I do not want them to become more dangerous. I do not want flooding and hailstorms and hurricanes to destroy communities and kill innocent people. I do want to make sure that all of our children and grandchildren and generations to come, both in the United States and around the globe, have the same opportunity to live well and prosper that I have had. I don't want us all to become poorer, as we spend increasing sums of money repairing the damage of climate change, damage that could have been prevented at far lower cost.[32] I don't believe it is fair for the profits of a few corporations to become the losses of us all. I believe that government is necessary, but I have no desire to expand it unnecessarily.

I also believe, as Pope Francis has stressed, that Earth is our common home and to disregard climate change is to disregard both nature and justice. As the Pope reminds us, his namesake was canonized because he "communed with all creation [and] felt called to care for all that exists."[33] Some would see such feelings of communion with Creation as at odds with cold-hearted scientific rationality, but in the eighteenth and nineteenth centuries it was a commonplace among European naturalists that scientific investigations were a means to come closer to God. In this, they were following the words in the Wisdom of Solomon, 13:5, which tells us that "through the greatness and the beauty of creatures one comes to know by analogy their maker." Or, as Haydn put it in his great oratorio of the late eighteenth century, the heavens are telling the glory of God—whatever we conceive that glory to be.

I also believe that history proves what John Donne wrote nearly four hundred years ago, that "any man's death diminishes me, because I am involved in mankind." I believe that I am my brother's keeper, and so are you. Is there not a reason, after all, that in Genesis the story that follows immediately on the heels of the Fall is the story of Cain and Abel?

The Old Testament—the foundation of the world's three great monotheistic religions—begins with Creation and so do the organizing myths and stories of most human societies. Whether we call it biodiversity or Creation or the Dreamtime or Mother Earth, climate change threatens it. Everything we know—from science, from history, from literature, from ethics—tells us that caring for our fellow citizen and caring for the environment are the same thing. The dichotomy of man versus environment, or jobs versus environment, or prosperity versus environment, is a dangerous fiction constructed to justify greed. It cynically warrants destruction in the name of the false prophet of progress.

That is what I believe.

If we fail to act on our scientific knowledge and it turns out to be right, people will suffer and the world will be diminished.

The evidence for this is overwhelming.[34] On the other hand, if we act on the available scientific conclusions and they turn out to be wrong, well, then, as the cartoonist says, we will have created a better world for nothing.

COMMENTS

Chapter 3

THE EPISTEMOLOGY
OF FROZEN PEAS

Innocence, Violence, and Everyday Trust in Twentieth-Century Science

Susan Lindee

Public anger, mistrust, and political retribution against members of the scientific community for what they believe or what they do are not new. What is new today is what I would call "scale." Scientific knowledge of direct and pressing practical importance is now being systematically rejected by some who stand at the pinnacle of global power, at a moment when the stakes are high for the survival of the planet. The scale of the problem of skepticism, doubt, and mistrust has escalated and taken new forms.[1]

It is at the level of scale that I wish to focus in comments here. I want to draw your attention "down" to a more intimate scale and to suggest that everyday technologies make visible the imbrication of science in quotidian life, and that they generate an ambient trust that should be generally recognized and nurtured.[2] I also propose that we have spent far too long not noticing or emphasizing this.[3] We have not drawn on everyday trust as a resource in public campaigns to restore trust in science more generally.[4] In their kitchens, up close, many people trust science implicitly, because it "works."

I realize that invoking the idea that "science works" is an unsatisfying theoretical response. As Oreskes's nuanced discussion of philosophy of science suggests, a proper, satisfying solution should operate closer to the rational realms of thought, presumably made manifest in a social and intellectual system that can evoke particular forms of consensus and therefore philosophical and scientific legitimacy. I agree, I agree, this is how things should be. We should trust science because it is the best we can do, and because it operates by relatively open and reliable social rules of evidence (intersubjective reason!) that have proven to be generally trustworthy, if not perfect, over the long haul, in many different ways. The efficacy and power of these ways of making knowledge is easy to see and for many of us requires no special pleading, no remarkable claims about moral order, no philosophical buttressing.

But in the populist spirit of our times, I suggest that instead we try to work our way up, from the toaster. The philosopher of science Nancy Cartwright has intermittently used the apparently simple and transparent technology of the everyday toaster to explore questions about causation.[5] Pressing a lever on a toaster, she notes, causes the bread to drop in to the toaster and begin to brown. This is a widely known phenomenon—perhaps you made toast yourself recently. But Cartwright suggests that assigning a cause to this browning involves accepting two "unrelated" causes—the lever itself, and the connection forged by its action to the electrical current that causes the heat. A toaster, Cartwright proposes, provides a way of thinking about and disaggregating causes. Following in her footsteps, so to speak, I want to propose in turn that the toaster and its everyday kin (frozen peas, iPhones, recycling bins, and even manufactured "wood look" bookshelves ...) also provide ways of thinking about trust and knowledge. I propose that everyday technology, widely trusted and even *loved,* can and should be re-scienced, made more *intellectual*—here in a country with such a long tradition of anti-intellectualism.[6] I propose this because I suspect that making more visible the deep presence of knowledge in everyday life could be another strategy to challenge public views of science as untrustworthy: In practice, whether people admit it, recognize it, and can articulate it, or not, everyday trust in science is ubiquitous, central to the taken-for-granted worlds of Twitter, refrigeration, air travel, and pharmaceuticals.[7]

Science is not just in the laboratory. It is everywhere and it is widely trusted and believed, I suspect sometimes by those who do not know that they are trusting science. We live in a physical world made of science and knowledge—we can't get out of it, even if we move to a rural enclave with organic goats and tomato gardens, or if we live on a cluster of palm-fringed islands in the Pacific, in a place once routinely subjected to nuclear tests. The Marshall Islands are already today inundated with rising sea-waters, and those rising waters, too, are scientific in more than one way.[8] Every day we move through systems built from scientific knowledge. We depend deeply on science that works and we traditionally do not think about this fact. We often casually "naturalize" the technological world derived from knowledge systems, in practice severing things like frozen peas from the systems of laboratory knowledge that make them possible. But the systems of knowledge implicated in frozen peas are vast, almost astonishing: Modern geological sciences in the oil and gas industry, the chemical development of plastics, scientific agriculture and the genetic modification of crops, chemical understandings of the freezing process, even the social sciences of marketing and persuasion. Frozen peas are saturated with reliable truth.

Many beloved and highly trusted technologies of everyday life are the direct result of legitimate and trustworthy scientific research. This simple fact is socially obscured and relentlessly disappeared in the boundary work around what counts as science and what counts as *technology*. While the terms are useful enough, their contemporary usage obscures important relationships between knowledge and practice—relationships that could be leveraged today to bring home to the general public the degree to which science is generally, almost universally trusted, even loved.

Why is the everyday presence of science in everyday life so often illegible, invisible, unremarked—or even understood as *irrelevant* to the question of whether scientists and science as an enterprise can be trusted?

Naturally I have a historical explanation. It hinges on the highly policed distinction between science and technology, and on why that distinction has been so important to the scientific community.

The sharp and exaggerated differences that are commonly recognized between science and technology are a historical invention, fostered during the Cold War to distance elite, pure, science from technologies of war. They have a much deeper history, but the (contaminating?) mobilization of science in World Wars I and II became a profound turning point. World War I was "the chemists' war," transformed by the German chemist and later Nobelist Fritz Haber's development of chemical weapons, and by the subsequent involvement of all forces in the use of a weapons technology later seen as illegitimate or even immoral. World War II was "the physicists' war," which ended with an application of "pure physics" that destroyed two Japanese cities, and led to arms races and arms-control efforts focused on a technology later seen as illegitimate and even immoral. For the scientific community in the United States and elsewhere, these technological achievements came to be seen as a profound threat to the legitimacy of science.

Critics of scientists' roles in both these wars included many leading scientists themselves, who expressed anguish about the roles of technical knowledge in state violence and about the actions of scientists themselves. After 1918 German chemists, blamed for the first use of chemical weapons, were excluded from scientific meetings and symposia for almost a decade.[9] Meanwhile with the rise of fascism in the 1930s, the international scientific community began a sustained philosophical construction of free scientific inquiry as a guarantee of healthy democracy, a uniquely pure endeavor.[10] Scientists' *moral* character was presented as proof that scientific inquiry could be trusted—thus locating trust in the religious or spiritual qualities of individuals. As Shapin has suggested, this was a proposal already obsolete: The rise of a modern professionalized scientific workforce after the 1880s, and the materialism implicit in Darwinian evolution by natural selection and other scientific ideas, had already coalesced to undermine the idea that science was a reliable path to the will and purposes of God, or that the scientist therefore occupied a moral position (because of this divine path-making labor).

By middle of the twentieth-century, earnest attention to the *nature of science* and its human dimensions was the focus of a compelling debate that can almost be seen as a reenchantment project (of science, not nature). Popular texts, such as Jacob Bronowski's *Science and Human Values* (1956), catalogued scientific virtue and construed the scientist as a uniquely moral actor—who was emphatically not responsible for the violence of modern warfare (or the atomic bomb). Scientists issued programmatic statements, like Vannevar Bush's *Science, the Endless*

Frontier, that expressed a view of science as benevolent, central to the welfare of mankind, and linked to the robust health of a functioning democracy. Meanwhile lavish defense funding for virtually every conceivable form of science—physical, biological, social—provoked fears that scientists were becoming intellectual slaves of the security state. How to safeguard an endeavor in which security clearance procedures could lead to lost jobs? The rhetoric around the purity of science became almost shrill.

And in these soaring narratives, technology occupied a distinctly different moral space—both less intellectual and less inherently moral—less *pure*. Technology was tangled up with social life and politics, messy, violent, dependent, and undeserving of the special status of science. Technology produced ruined cities and irradiated people. Science was doing something else.

The discipline of the history of science itself emerged as a respectable academic discipline—with faculty appointments at multiple universities—during this midcentury moment of the promotion of science's massive and unbridgeable distance from impure technology. Many historians of science in the 1950s took it as their mission to reinforce public faith in pure science as a bulwark against fascism and communism. In the 1950s historians told scholarly stories about autonomy and purity, even as all around them at their home institutions the living scientific community wrestled with lack of autonomy, new forms of nationalism, and a security state that enforced political conformity in draconian ways. Historians in this early efflorescence of the field engaged in intense boundary work, avoiding attention to (mere) technology, pseudoscience, or folk knowledge of any kind. For example the idea that alchemy was *not science* led to its complete erasure for decades from the life of the remarkable natural philosopher Isaac Newton—who himself as things turned out cared quite a bit about alchemy (in 1975, Betty Jo Teeter Dobbs brilliantly demonstrated the importance of alchemy to Newton's thought).

Even more revealing, in the summer of 1957 the president of the History of Science Society Henry Guerlac refused, at an infamous meeting, to even consider opening the society meetings or journal to scholars interested in the history of technology. As it was described in a 1998 appreciation of historian of technology Melvin Kranzberg, "Guerlac's refusal to allow either, lest the history of science be tainted by such alleged intellectual inferiority—by attention to lowly 'tinkerers' rather than great 'thinkers'—is what spurred Mel [Kranzberg] to establish a separate society with its own journal. The experience reinforced his growing conviction that the history of technology required and deserved autonomy as an intellectual endeavor."[11]

Thus emerged a sharp, highly policed line between pure science and impure technology. Both terms seemed to reference something grand and self-evident: Scientists made knowledge, technology was *only* knowledge applied. This idea placed those physicists who made the bomb in the arena of truth of nature—with their theories of atomic structure—and allocated the actual weapon itself to engineers, who were beneath historical attention (for historians of science). Biological weapons, chemical weapons, and ICBMs were all "technology," not science. Thus the idea of a clear and inviolable distinction between science and technology was promoted in ways that placed science in a morally privileged position.

Indeed, the Johns Hopkins geophysicist Merle Tuve, who led the effort to develop a new proximity fuse in 1941 at the Applied Physics Laboratory, saw the distinction between science and technology as a central part of his own scientific identity. In her 2012 paper, Wang has beautifully characterized the logic and practice of this idea in Tuve's professional life. As Tuve told an audience in 1958, "Science is not airplanes and missiles and radars and atomic power, nor is it the Salk vaccine or cancer chemotherapy or anticoagulants for heart patients. These all are technological developments. . . . Science is knowledge of the natural world about us . . . it is the search for new knowledge about the marvelous world in which we find ourselves."[12] Note that Tuve even left out the Salk vaccine!

We can sympathize with his desire to draw this line—he was a scientist who helped make weaponry—but we need to recognize its historical specificity, and its limitations as a complete explanation of the relevant relationships between different forms and iterations of technical knowledge. It is also important to understand that the forces and tensions that animated this boundary work were deep and oppressive for those who managed professional lives as scientists in the Cold War in the United States. The stakes were very high for scientific professionals.

In a heartfelt July 1954 letter to a powerful Atomic Energy Commissioner, for example, the Yale University biophysicist Ernest Pollard described how he learned to keep secrets. "Many of us scientists learned the meaning of secrecy and the discretion that goes with it during the war," Pollard said. "We had very little instruction from outside." When the war was over, he made a conscious decision to avoid secret research. He "thought carefully through the problems of secrecy and security" and made the decision to handle only material that was entirely open. "I returned one or two documents I received concerning the formation of the Brookhaven Laboratory, in which I played a small part, without opening them." But the outbreak of the Korean War, in June 1950, and his own concerns about the Soviet Union, led to a change of heart. He came to feel that "I as a scientist should pay a tax of twenty percent of my time to do work that would definitely aid the military strength of the United States."[13]

In the process, as he engaged in secret research during the Cold War, he learned a form of extreme social discipline that he called "the scientists morality." "I have learned to guard myself at all times, at home, among my family, with the fellows of my college when they spend convivial evenings, with students after class asking me questions about newspaper articles, on railroad trains and even in church. It has been a major effort on my part, unrelenting, continually with me, to guard the secrets that I may carry."[14]

Pollard's comments resonate with those of many other experts in the heart of the Cold War in the United States. Being a scientist often meant concealing one's work and ideas from friends, family, students, colleagues. An enterprise founded on an ideology of openness and free exchange became increasingly oriented around keeping secrets.[15] Individual scientists could lose their jobs if they lost their security clearances.[16] And security clearance could be withdrawn for a wide range of infractions, including accepting dinner invitations from people who were members of the Communist Party.[17]

Scientists even lost jobs for refusing to testify when called before the House Committee on Un-American Activities by Senator Joseph McCarthy (R-Wisconsin).[18] The physicist David Bohm, who lost his assistant professor job at Princeton for this reason, went on to make illustrious scientific and philosophical contributions under difficult circumstances in Brazil and later, in the United Kingdom.[19]

Scientists were also harassed by the 1950s equivalent of Twitter trolls. Henry deWolf Smyth, who voted to permit the Princeton physicist J. Robert Oppenheimer to keep his security clearance (seen by some as an unpatriotic act), received a threatening letter from an "Angry American Family" who promised "some day we Americans will catch up to all of you traitors."[20] The geneticist Arthur Steinberg was also subject to shocking public attacks, losing a deal on a house and several jobs because of inaccurate reports that he was a Communist. Steinberg gave only thirty-five documents to the historical archives at the American Philosophical Society (APS) in Philadelphia. All of them chronicled the cruelty with which he was treated after being accused.[21] A January 1954 letter from his attorney to a housing development where Steinberg and his wife had attempted to purchase a home described "the anonymous phone calls which my client received from some neighbors threatening dire consequences if they lived in the house." A 1948 letter from a colleague openly stated that Steinberg had been removed from the list of viable candidates for a job, because departmental faculty had heard about "the Communist charges." In selecting what he chose to donate to APS archives, Steinberg clearly intended that his painful experiences not be forgotten.[22]

Other scientists made quiet bargains, parsing out their time, so that some percentage of their professional life went to "pure science," and some to defense-related work in the name of patriotism. Like Pollard, they drew the boundaries of their professional lives with personal calibrations of responsibility, with many kinds of lines in the sand. MIT biologist Salvador Luria announced in 1967 that he would not work on *any* defense projects in protest against the Vietnam War.[23] More judiciously, in 1969, Ronald F. Probstein and his fellow researchers at MIT's Fluid Mechanics Laboratory made what they called a "directed effort to change" their research. They reduced the amount of military-sponsored research they were doing from 100% to 35%, with the remaining 65% explicitly devoted to "socially oriented research." The point was not to sever all ties to military research. Rather, they wanted to "redress an imbalance."[24]

These experiences shed some light on the struggles and strategies of the rank-and-file experts who fueled economic growth and facilitated national defense in the heart of the Cold War in the United States. They learned to keep secrets, lie, and pass polygraphs. They shared tips about what to say in security clearance hearings, how to burn trash, managing selective service requirements, concealing the military relevance of a project, and managing the anger of their peers. They became vulnerable to science swerved by defense interests, to possible prosecution or fines, even deportation, and to the skepticism of their peers either because their peers believed them to be disloyal, Communists or socialists, or because their peers viewed them as overly dependent on defense funding. They also learned how to make things and ideas that produced massive human injury. The professional and personal stakes were high; the risks real, the embeddedness of knowledge in the state's monopoly on violence profound.[25] Within the scientific community, who could be trusted, both as an expert witness to nature's ways, and as a proper patriot, on the right side of a global ideological war? And in the broader civic world of political and social order, how could trust in the scientific community be sustained, when scientists made bombs, concealed from the public the nature of their work, and turned on each other in such a toxic political environment? Secrecy does not usually engender trust. And infighting within a community can undermine its legitimacy.

Oreskes suggests that science can be trusted because flawed scientific ideas in the past were subject to contemporaneous criticism—there were individuals objecting to scientific theories later found to be inaccurate, for example about women's bodies, eugenics, or plate tectonics. Their existence—their voiced public objections—exemplify in some way the self-correcting properties of scientific knowledge, she proposes. In practical terms, this might not be as powerful an argument as it has long been presumed to be. And more ominously, minority objections from people who hold PhDs to climate science, evolution, and other scientific ideas are not unheard of today. Indeed, it is entirely possible to find PhDs in astronomy who have believed that aliens came to earth some time ago, and now live among us.[26] Singular voices of any kind are not necessarily reassuring.

The activities of the "alt-science" advocate Art Robinson, a trained PhD who once collaborated with Nobel Prize winner Linus Pauling (and who seems to have made a career out of proclaiming that Pauling was "wrong"), suggest just how diverse the scientific community is in practice.[27] The existence of diverse voices means only that the credentialed community of knowledge producers can and does include people who think very differently. What bearing do such differences of viewpoint and opinion have on the legitimacy of public trust in science? Oreskes places consensus at the heart of her analysis, proposing that science is a collective accomplishment, and that this collectivity is the source of its legitimacy and power. It comes into being through a process, with twists and turns, dead-ends, disputes, and resolutions, and the messiness of this process is a virtue, rather than a flaw. As she tracks the disturbing current state of affairs, Oreskes shows how much science now needs defenders, and defenses. We may have lost, she proposes, the Enlightenment vision of trustworthy natural knowledge that could be made by human (and inherently flawed) actors through disciplined rules of testing and experiment. But we still retain the power of consensus and reason. This kind of argument is utterly persuasive to me. It may not be to some of those most determined to disbelieve the findings of evolution or climate change or vaccine science. And yet many people do trust science. They may not fully recognize that deep trust, because science has been so resolutely distinguished from technology, for so long, with such intensity. It is possible that most citizens in industrialized societies have a kind of subterranean, unreflective trust in technical knowledge because "it works" and because it is, almost literally, the texture that defines their lives, every hour of every day. One of the struggles of all social theory is to find a perspective from which the waves and gravity can be detected, the water we swim in experienced—a problem Einsteinian in its dimensions. Where should we stand to understand the problem of trust? What are the right questions?

As the science studies scholar Donna Haraway put it in a famous 1988 paper, "situated knowledge" reflects the politics and epistemologies of location (in every sense of the term). Science makes "claims on people's lives" and there should be room for views of nature in which "partiality and not universality" is seen as rational. The "view from a body," she proposed, is paradoxically more powerful than "the view from above, from nowhere, from simplicity."[28] Haraway was responding to the difficulties feminist scholarship faced in relation to "objectivity." Some feminist scholarship at the time seemed to be engaged in a brutal "unveiling" project, which would demonstrate that scientists were irretrievably biased, that knowledge (for example knowledge supporting the idea of female inferiority) was contaminated by social beliefs, and that even (most terrifying) perhaps there was *no* objective knowledge at all, no position from which truth could be seen. These forms of "high" social constructivism, in which knowledge was reduced to social interests and hopelessly contaminated, threatened to reduce technical knowledge to irrelevance, a threat that Haraway and many other scholars found disturbing. How could feminist theory facilitate an understanding of a real world that could be friendly to human needs if it rejected the possibility of legitimate knowledge?[29] Feminist objectivity, she proposed, "makes room for the surprises and ironies in the heart of all knowledge production; we are not in charge of the world."

As recent events have made clear, the scientific community is decidedly "not in charge of the world." For at least a century, it has wrestled with the growing relevance of virtually all scientific fields to the "garrison state." Now it finds a new kind of embeddedness, in a different kind of state power. The experiences of scientists in the Cold War, as they navigated totalizing systems of political control, have a new, chilling resonance. Their situations—their positions and embeddedness—help us see some of the ironies of knowledge production. None of the scientists I consider in this essay had the option of opting out completely. It was not possible to do so and to continue to engage in scientific labor. Even if they did not do defense work, they were training students who would. How, then, did they make sense of their predicament? And what can their predicaments teach us about the predicaments of scientists today?

The generation I focus on had learned in the course of their formal education in the 1930s that science was open, universalistic, internationalistic, and an endeavor focused on the "welfare of mankind." Yet in practice, in the heart of the Cold War, for many scientists, their research was not open but secret, not internationalistic but nationalistic, and not conducive to welfare but engaged with the sophisticated technical production of injury to human beings. These forms of injury were realized through new weapons, new surveillance methods, new information systems, even new ways to interrogate prisoners using psychological insights, bring down economies, or start epidemics—"public health in reverse"—biological weapons. Experts in fields from physics to sociology found their research calibrated to empower the state, and scientists trained to see themselves as creating knowledge as a social good found themselves engaged in something that felt very different *to them*. Professional societies from the American Association for the Advancement of Science, to the American Society of Microbiology, to the American Chemical Society created committees on "social issues" and produced statements on science and the "welfare of mankind" through the 1950s, '60s, and '70s. Meanwhile their members made weapons and worked in the defense industry.

The profound struggles around science and violence in the twentieth century animated public strategies that corralled science, and moved it safely out of the kitchen, the clinic, the urban street, or most importantly, the battlefield. By drawing sharp boundaries between pure and applied knowledge, many scientists pursued a sometimes morally encoded strategy intended to preserve the pure core of technical truth, *the innocence of pure science*, by isolating it from *technological* things like guns, gunpowder, bombsights, nuclear weapons, chemical weapons, and psychological warfare. This strategy often enforced a hierarchy separating scientists from engineers, physicians, and other experts who "got their hands dirty."

I would suggest that it may also have played a role in obscuring from general view the saturation of everyday life today with scientific knowledge.

Oreskes has called us to think critically about public trust in science. It is clear that the answer is not that science should be trusted because it is always true, right, accurate. It is not always any of those things. But it is trustworthy despite its humanness, its vulnerabilities to misunderstanding, error, misplaced faith, social bias, and so on. It is trustworthy because of the sustained human labor that goes into making it, the integrity of the process, and because it has already transformed human life in so many ways that are obvious, transparent, profound.

All of the protocols that Oreskes describes are conducive but not determinative of validity. She invites us to think about the fragility of knowledge as a resource for trust, suggesting that people should trust science precisely because it is a system responsive to evidence, observation, experience. This is a stronger argument than perhaps even she realizes, for when we add everyday technology to this potentially trust-generating mix, we find something familiar that can be persuasive. For many people the reliability of everyday technologies might be as close as they ever get to scientific knowledge. But it is closer than it seems.

In his *New Yorker* review of the historian of technology David Edgerton's book *The Shock of the Old*, the historian of science Steven Shapin described himself writing in his kitchen, where he was surrounded by technology: a cordless phone, a microwave oven, and a high-end refrigerator, while working on a laptop. His essay noted that "the texture of our lives would be unrecognizable" without these things made from technical knowledge.[30] His comments were inspired by Edgerton's provocative book, in which he explores the "creole" technologies of what he calls, bluntly, the "poor world."[31] Creole is a term that commonly refers to local derivatives of something from elsewhere—such as cars from 1950s Detroit that are still running in Havana. It is not generally intended to suggest sophistication, innovation, or elite knowledge. But Oreskes shows us that elite science too has properties of "creole" making do, cobbling together data and ideas that can stay on the road, without being perfect and without the advantage of the original parts. It is not an idealized social and intellectual system of pure truth, free of misunderstanding, confusion, or error. It is pretty much the best we can do, and much of time, it works—like that iPhone.

We swim in science every hour of every day, but we don't talk much about it. The many roles of science in our lives are naturalized or black boxed. Many of those who question climate change or vaccines are more than happy to deploy drones as technologies of war, or for that matter to use Twitter. Drones depend historically on layered and clustered types of scientific theory and practice, going back many decades. Meanwhile those who promote creationism share their ideas on the web, which is a result of defense support for scientific research—in mathematics, electromagnetics, physics, and other fields. So both trust and mistrust of science are discriminating, selective, and biased. Political leadership in the United States controls the most powerful military in the world—power built by scientists and engineers. Yet this achievement of high technocratic reason, this weapons system that truly "works," in fairly spectacular terms, does not seem to confer legitimacy on the enterprise of science-in-general.

The late Carl Sagan is haunting me lately so I will quote him: "We live in a society exquisitely dependent on science and technology, in which hardly anyone knows anything about science and technology." I agree with his claim that we live in a society that is exquisitely dependent on science and technology. But I am less sure about his claim that hardly anyone knows anything about it. People know the things that live inside their everyday lives, but they emphatically don't see them as scientific.

So, maybe we need an epistemology of frozen peas. For frozen peas, if interrogated historically, involve as many layers of science as do drones. As I have already noted, these layers include the geology of oil exploration, the development of monomer chemistry and polymerization, the impact of the evolutionary synthesis on agricultural breeding, the development of the new genetics and GMOs, the scientific understanding of conservation of matter in temperature change, developments in bacteriology, and even knowledge from the social sciences, in psychological theories of marketing, imagery, and persuasion that have reshaped the consumer experience in the twentieth century—helping manufacturers understand for example how to persuade people to buy and eat frozen peas.[32]

You might not pay much attention to frozen peas, and I can't blame you, but I use this commonplace and seemingly simple food technology to suggest just how invisible this scientific world-making has become—almost as though it were structured to be invisible. Scientists long ago began to systematically distance themselves from the technologies that their insights and understandings could produce. And their distancing has been successful. But people love and trust technology. The flow of prestige and legitimacy "down" from science to technology—the flow of trust, viability, proof of value, of "working"—should perhaps be transposed, for the good of science and the good of the world.

Chapter 4

WHAT WOULD REASONS FOR TRUSTING SCIENCE BE?

Marc Lange

The question of why we should trust science can easily induce a kind of dizziness and even despair. Suppose we try to argue that we should trust science because rigorous application of the scientific method has generally led to successful results: the discovery of many truths and the rejection of many falsehoods. This approach will not get us very far. It invites the reply: What is our basis for judging that the current verdicts of science consist of many truths? Apparently, our basis is that the current verdicts of science line up with our *beliefs* about what is true. But if (as is likely) we have arrived at our beliefs by relying on science, then we have only traversed a very small circle; we have tried to use science to vindicate itself, which cannot succeed.

This kind of circularity is difficult to avoid. Suppose we try to argue that we should trust science because science has led to many accurate predictions and technological achievements. It has allowed us to predict accurately the outcome of adopting certain public health measures or of putting batteries and wires into certain configurations. As Professor Oreskes mentioned, the track record of science in these matters is rather good. So, the argument would conclude, we should trust science.

But this reasoning was *itself* an instance of using science. The reasoning began by noting a pattern in our experience so far (that science has worked pretty well in the past). The reasoning then took these data as good evidence that the pattern will continue into the future. This is a good example of *scientific* reasoning. But to use scientific reasoning seems to beg the question, since our goal was to justify putting our confidence in scientific reasoning in the first place. (Thus there may be some concern about Comte's suggestion, mentioned by Professor Oreskes, that we support science by studying scientists scientifically.)

The same charge of circularity could be lodged against the idea, mentioned by Professor Oreskes, that we should especially trust science when science has employed peer review and other agreed-upon scientific practices, or that we should especially trust science when we are dealing with the verdict of recognized scientific experts working within their recognized field of expertise and conforming to recognized scientific procedures. Who is to recognize them? On what basis is someone to be judged an expert? By the endorsement of other experts, each of whom gets his or her status by the endorsement of still other experts? As I said, it is easy to succumb to a kind of dizziness at this point, with the looming threat of infinite regress or vicious circularity. It is also easy to imagine the American Enterprise Institute or young Earth creationists criticizing this incestuous pattern of experts vouching for other experts.

This kind of extremely corrosive doubt has a venerable pedigree in philosophy. It can be traced back to David Hume[1] in the eighteenth century, who is generally credited with having posed "the problem of induction." ("Induction" denotes the form of reasoning by which hypotheses are confirmed by evidence in science.) Such pervasive doubt can be traced back even further—to René Descartes[2] in the seventeenth century and to Sextus Empiricus in the first century, who famously wrote:

> Those who claim for themselves to judge the truth are bound to possess a *criterion* of truth. This criterion, then, either is *without* a judge's approval or *has* been approved. But if it is *without* approval, whence comes it that it is truthworthy? . . . And, if it *has* been approved, that which approves it, in turn, either has been approved or has *not* been approved, and so on ad infinitum.[3]

There are at least two recipes for combatting the dizziness and despair that this kind of wholesale skepticism tends to provoke. One recipe is to point out that to ask for a justification for science *as a whole* is to make an unreasonable demand. It is like asking someone to justify trusting in reason as opposed to faith, wishful thinking, or astrology. If you give a *reason* for trusting in reason, then you have presupposed what you are trying to show. On the other hand, if you give something that is *not* a reason for trusting in reason, then that is giving no sort of justification at all. The game is rigged. You should simply reject the demand to offer a reason when there is nothing that could possibly, even in principle, count as one. As Professor Oreskes mentioned, one of the most important features of science is that it is self-correcting; it is able to put in jeopardy *any* of its theories, scrutinizing its justification. But science *cannot* reasonably be expected to put *all* of its theories in jeopardy *at once*.[4]

There is a second important recipe for combatting the dizziness and despair of the skepticism that we have been discussing. The question of why we should trust science might be put this way. Scientists begin by making a bunch of observations. That is supposed to be the first stage of scientific research—at least, first in logical order. In the second stage, scientists use their observations to confirm various theories that make predictions about what would be observed under various conditions—or perhaps to confirm various theories that purport to reveal the unobservable causes or mechanisms responsible for what we have observed. What justifies this second stage—this inductive *leap* from the safe ground of past observations to the risky business of predicting future observations or positing hidden mechanisms? The history of philosophy is littered with attempts to support this second stage given the first. But most of these attempts have ultimately been judged to be question-begging in one way or another.[5]

If the challenge is to justify this second stage of scientific research *given the first stage*, then we can respond with the second recipe that I want to mention for combatting skeptical dizziness. The recipe is to point out that even at the first stage, we are already taking for granted that we are justified in engaging in the second stage; we are already presupposing that certain steps beyond our observations are justified. After all, if I take myself to be observing something to be the case, then I must believe that I am *qualified* to make that observation—that by dint of my training, I am now able to tell that something is the case simply by looking or hearing or smelling or whatever. I am not purporting to be infallible, of course. But I am purporting to be reliable enough that I am worthy of trust (in the absence of some specific reason to doubt my accuracy) in this case. Without this additional belief about myself, I cannot justly take myself to have genuinely observed that something is so.

But how did I become *justified* in this additional belief about myself? The justification must derive from my own history of having purportedly observed things. My *past* behavior of saying that I saw various things, nearly always and only when those things were in fact there, justifies my belief that I will typically be reliable *in the future* when I say that I have seen those things. A generalization like that obviously goes beyond what we have already observed. So in making an observation, we must already allow that we are justified in believing various generalizations about matters that go beyond what we have already observed. As the philosopher Wilfrid Sellars says, in arguing for a similar point:

> The classical "fiction" of an inductive leap which takes its point of departure from an observation base undefiled by any notion as to how things hang together is not a fiction but an absurdity.... [T]here is no such thing as the problem of induction if one means by this a problem of how to justify the leap from the safe ground of the mere description of particular situations, to the problematic heights of asserting lawlike sentences and offering explanations.[6]

What I have been discussing so far is the demand for a reason for trusting science *as a whole*, a demand that is problematic precisely because its target is science as a whole. This kind of *wholesale* demand for justification should be distinguished from a *retail* demand that asks why some *particular* scientific result should be trusted. That kind of question, unlike the wholesale question, *can* be answered without circularity—by appealing to other scientific findings. This is the approach implicit in the series of case studies that Professor Oreskes offers. In each of those cases, participants were called upon to present the epistemic credentials of various scientific hypotheses. In none of those cases were participants called upon to underwrite science as a whole.[7]

However, a source of trouble for this wholesale/retail distinction comes from cases where very large bodies of theory are being called into question. As Professor Oreskes mentioned, the philosopher Thomas Kuhn[8] was very influential in questioning the rationality of what he termed "scientific revolutions," where an entire "paradigm" is called into question. A scientific revolution, according to Kuhn, is (in the phrase that Kuhn popularized) a "paradigm shift." Rival paradigms disagree on what sorts of facts scientists can directly observe, what sorts of measuring devices are reliable, and how to interpret criteria of theory choice such as simplicity, accuracy to observations, explanatory power, and fruitfulness. As Professor Oreskes mentioned, Kuhn regarded rival paradigms as sharing only the most minimal standards for theory choice. So in a crisis, when the prevailing paradigm is in jeopardy and a justification is demanded for one of the new candidates for paradigm (or for sticking with the incumbent), the common neutral ground arbitrating among the rival candidates is too meager to sustain any powerful reasons favoring one candidate over another. This is one aspect of what Kuhn calls "the incommensurability of paradigms," as Professor Oreskes mentioned.

In a crisis, then, a *retail* challenge to a *particular* scientific theory becomes a *wholesale* challenge to the results of an entire scientific field. What sort of noncircular justification can there then be for one paradigm over another when scientific *methods* and scientific *theories* interpenetrate? In other words, Kuhn argued very successfully against a sharp boundary between scientific *methods* and scientific *theories*. What methods scientists believe to be reliable are informed by what scientists believe about what the world is like. But if theories and methods interpenetrate, then as scientific theories change, scientific methods change. There is then no permanent, neutral method to arbitrate between rival paradigms; although their common ground presumably includes mathematics, deductive logic, and the probability calculus, that is not enough common ground to serve as a neutral arbiter to decide the issue. This is the challenge that Kuhn launches at us.

One promising reply to Kuhn's challenge is to acknowledge that deductive logic, arithmetic, the probability calculus, and whatever else is permanently neutral common ground in science is *not* enough to decide in a crisis among the rival candidates for paradigm. Nevertheless, in a *given* crisis, there will be more common ground among the rivals than merely that which is common ground across *every* crisis. In different crises, this additional common ground will be different. Although it may be meager, creative scientists in various crises have found ways to take the common ground and extract from it powerful reasons supporting one candidate for paradigm over another.

Let us look briefly at an example. Galileo managed to find a way to generate a powerful argument from the meager common ground available in the crisis of terrestrial physics occurring during his time. Galileo proposed that if a body falls freely to Earth from rest, then in each succeeding interval of time, the distance covered by the body grows as the odd numbers, so if 1s is the distance that the body traverses in the first time interval, then in the succeeding intervals, it covers the distances 3s, 5s, 7s, 9s, 11s. . . .[9] Other scientists proposed rivals to Galileo's "odd-number rule." Honoré Fabri proposed that the distances traversed grow as the natural numbers (that is: 1s, 2s, 3s, 4s, 5s, 6s . . .). Pierre Le Cazre proposed that the distances grow as the powers of 2 (that is: 1s, 2s, 4s, 8s, 16s, 32s . . .). An *experimental* argument favoring one of these theories against the others would have been ineffective as long as there was no shared paradigm for terrestrial physics and so no agreement on which measuring devices were accurate for measuring time and distance and on when to blame disturbing factors (such as air resistance) for a theory's failure to perfectly match with observation.

Nevertheless, according to a 1627 letter from Gianbattista Baliani to Benedetto Castelli, Galileo had a powerful argument favoring *his* proposal over these rivals.[10] His argument was that if one of these rivals holds for time intervals expressed in one particular unit (say, in seconds), then that proposal will not hold for time intervals expressed in another unit (say, in minutes). If we take the distances given by these proposals for successive one-unit intervals

> Fabri: 1s, 2s, 3s, 4s, 5s, 6s . . .
> Le Cazre: 1s, 2s, 4s, 8s, 16s, 32s . . .

and switch to a new unit of time that is twice as long as the original, then we find the distances fallen in these new intervals to be

> Fabri: 3s (= 1s + 2s), 7s (= 3s + 4s), 11s (= 5s + 6s) . . .
> Le Cazre: 3s (= 1s + 2s), 12s (= 4s + 8s), 48s (= 16s + 32s) . . .

These distances do not fit the proposals. For example, if the proposal is that the distances grow as the natural numbers, then that proposal is violated when we switch units since the distances 3s, 7s, and 11s do not stand in the ratio of 1 to 2 to 3.

Thus, if one of these rivals to Galileo's proposal holds in one system of units, then it will not hold in certain other systems of units. Galileo's point, expressed in today's terminology, is that neither rival to his proposal is "dimensionally homogeneous." Roughly speaking, we can define "dimensional homogeneity" as follows:

> Relation R is "dimensionally homogeneous" exactly when it is a broadly logical truth that if R holds in one system of units, then R holds in any system of units for the various fundamental dimensions (e.g., length, mass, time) of the quantities so related.[11]

Of course, a relation can hold without being dimensionally homogeneous. For example, on a given date it may be that my son's weight equals my age—but this relation holds only if my son's weight is measured in pounds and my age is measured in years. The relation is therefore not between my son's weight and my age *themselves*, but rather between their measures *in a particular system of units*.

Part of the meager tacit background belief common to all parties during the crisis of terrestrial physics in Galileo's time was that the relation among the distances traversed in successive equal time intervals by a body falling freely from rest is *independent* of the unit of measure. It is a relation among the distances *themselves*, not between their measures in some particular privileged unit. Presumably, neither Fabri nor Le Cazre ever bothered to say explicitly that they believed the relevant relation to be dimensionally homogeneous. But neither had to; it was understood. As far as I know, neither specified some particular units as having to be used to measure distance or time. Galileo argued that since neither of these proposals is dimensionally homogeneous, the relation in question cannot be given by either of them.

By contrast, Galileo's own proposal *is* dimensionally homogeneous. If we take the distances it dictates (1s, 3s, 5s, 7s, 9s, 11s . . .) and switch to a unit of time that is twice as long, we find that, by Galileo's proposal, the distances covered in these longer intervals are 4s (= 1s + 3s), 12s (= 5s + 7s), 20s (= 9s + 11s). . . . The ratio of 4 to 12 to 20 is the ratio of 1 to 3 to 5—the odd-number sequence that Galileo's rule demands.[12]

That there is a dimensionally homogeneous relation among the free-fall distances was tacit common ground in this *particular* crisis, but it need not be common ground in every crisis. Perhaps only a scientist with Galileo's ingenuity would have found a way to turn such sparse common ground into a strong argument for one of the rival theories against the others. Finding powerful reasons in a crisis is inevitably going to be difficult. But it is not impossible. (I think that we could take this lesson to heart in many disputes, even outside of science.)

I will conclude with one further point. According to a 2012 Gallup poll, 46% of Americans deny the evolutionary origins of human beings.[13] In that same year, a congressman on the House *Science* Committee referred to evolution and the Big Bang as "lies from the pit of Hell."[14] Whether an American believes that climate change is taking place is highly correlated with that American's party political affiliation.[15] In this kind of political climate, I think we need to do a better job of communicating the rational basis of science. In our classes, we philosophers love to break things; we love to present the failures of various venerable proposals regarding the logic of scientific reasoning. But we need to go beyond incommensurability and underdetermination and the Duhem-Quine thesis and the new riddle of induction[16] and the demise of the demarcation problem[17] and the pessimistic historical meta-induction.[18] Where we can, we need to give a *positive* account of the logic underlying scientific reasoning. Our students are able to grasp a positive account and they are hungry to see one. We owe it to them—and to ourselves—to supply one.

Chapter 5

PASCAL'S WAGER
REFRAMED

Toward Trustworthy Climate Policy
Assessments for Risk Societies

Ottmar Edenhofer and Martin Kowarsch

The Trump administration seems to accept climate science, though it argues against ambitious climate change mitigation efforts. In October 2017, the US Environmental Protection Agency (EPA) made a remarkable new proposal as to how climate change impacts can be evaluated in economic terms, i.e., how to calculate the "social costs of carbon" (SCC).[1] They suggested social costs of only $1–6 per additional ton of carbon dioxide (CO_2) emitted to the atmosphere in the near future. These numbers are extremely optimistic compared to the $45 per ton of CO_2 estimated under Obama's presidency. In contrast to the Obama administration, which included the global damages of climate change in its estimate, the new EPA calculations only take into account US domestic damages. If policy makers and investors base their decisions on the newly suggested numbers, ambitious US climate policy can hardly be justified. Calculating such costs does presuppose basic trust in the underlying climate science. This example illustrates a thorny point: scientific consensus does not imply policy consensus. Instead, calculating SSC implies controversial value judgments. Hence, in Trump's universe, climate science does not commit the United States to ambitious climate policy, the provision of public goods, or far-reaching international cooperation. Rather, by virtue of heavily discounting future generations and dismissing public goods as globalist twaddle, the United States does no longer seem bound by its Obama-era commitments. Since scientific consensus on human-induced climate change still allows for a multiplicity of "value vectors" and policy pathways, a serious analysis of the entanglement of facts and values in expert studies, the role of scientific expertise, and the design of policies is direly needed in risk societies—i.e., societies that have to deal "with hazards and insecurities induced and introduced by modernisation itself."[2]

Naomi Oreskes makes a compelling case for the trustworthiness of science, its social, value-laden, and fallible character, and the conditions for objectivity. She also reveals the rather effective diversion created in public debates by many climate change skeptics. In most cases, skeptics no longer question the overwhelming, credible scientific evidence on anthropogenic, risky climate change. Instead, they argue against particular economic or social implications of potential political *responses* to climate change. While they balk at the idea of engaging in open dialogue about their concerns as to the various possible (political) responses to climate change, they simply deny that the latter is a scientific problem. They urge the academic community to continuously clarify the science. And even when the science of climate change is accepted they deny the severity of damages caused by climate change. The recent SCC estimate of the Trump administration is one such example. Furthermore, according to their view, evidence-based research can never reduce the uncertainties to such an extent as to allow for the implementation of climate policies.

The often-intractable conflicts about climate policy are thus not necessarily rooted in a lack of trust in climate science—but rather in disagreement on the design of climate policies.[3] Any reasonable answer to this crucial question is dependent on disputed ethical values, concepts of intergenerational and intragenerational justice, preferences, and interests (as Oreskes rightly points out). But the estimates of social science concerning the costs of action and inaction, the co-benefits and unintended societal side effects of climate policy must figure prominently too in such an answer. Natural science and technology alone cannot determine appropriate climate policies. Stubbornly insisting on the facts of climate science to counter skepticism just makes environmental controversies worse.[4] Due to her emphasis on criteria for the trustworthiness of natural science Oreskes does not answer sufficiently the question as to the conditions—if at all[5]—under which the policy assessments provided by scientific experts can both be trustworthy and legitimate, particularly given the highly disputed value judgments involved. Therefore, we need to reflect about these integrated, transdisciplinary assessments of complex socioeconomic and political aspects. How can we specify, apply, and perhaps amend Oreskes's convincing criteria for trustworthiness to provide guidance also for the social sciences in these cases?

Oreskes does not highlight the indeterminateness of climate policy. Rather she presents a simplified version of Pascal's Wager[6] as applied to climate policy (see table 1). Oreskes suggests that ambitious climate policy would be beneficial even if the science supporting anthropogenic climate change turned out to be wrong. A risk-neutral decision-maker would choose ambitious climate policy, if $p > C / (V\text{-}E)$. In other words, ambitious climate

TABLE 1. Pascal's Wager Applied to Climate Policy

	Dangerous Climate Change Probability p	Harmless Climate Change Probability 1−p
Ambitious Climate Policy	Low Damages (E) + Mitigation Costs (C)	Mitigation Costs (C)
No Climate Policy	Irreversible High Damages (V)	Zero Net Costs (0)

policy is a no-regret option irrespective of the probability of climate change if the mitigation costs are negative due to the co-benefits of climate action. Such benefits include reduced local air pollution after phasing out coal or less dependency on fossil fuel imports. In contrast to Oreskes, the Trump administration has discounted the damages (V) to such an extent that investments in ambitious climate policy cannot be evaluated as beneficial for US society, even if the probability of dangerous climate change is very high.

The payoffs in the climate policy wager are not merely givens that result from a state of nature—they should be understood as an emergent outcome of a social learning process. Most experts would evaluate Oreskes's no-regret argument for ambitious climate policy as overly optimistic.[7] It is, however, equally pessimistic for Trump to ignore the damages of climate change for humankind as a whole by discounting future climate damages to an excessive degree—particularly when their partly irreversible character is taken into account. Rational decision-makers would choose to reduce emissions both immediately and considerably. They would launch a societal learning process designed to adapt climate policies according to new insights about future damages and mitigation costs as well as other effects and risks associated with these policies. Such a social learning process would draw on evidence from ex-post policy analysis to determine which policy instruments worked and which did not. Recent analyses of the EU ETS[8] are examples of such studies. The iterative nature of the learning process is conducive to making incremental, yet successful steps in climate policy on various governance levels while avoiding irreversible lock-in effects. If democratic societies want to tackle climate change as a serious risk for current and future societies they need to engage in rational discussions and learning processes about alternative, available solutions and their (often uncertain) implications. Such discussions must include an examination of the specific risks and the pros and cons of different policy pathways as well as interdependencies with different policy fields, governance levels, and time scales. A scenario in which human-induced climate change transpires to be a mistaken hypothesis would then be only one of many different possible scenarios—though a rather extreme and unlikely one, as Oreskes herself suggests.

Salient examples of difficult climate policy issues include: the evaluation and comparison of climate damages in different regions; appropriate prices on CO_2 emissions for different countries; the sector(s) that should be affected by a carbon pricing scheme; the specific distributional effects of carbon pricing on different social groups; policies that could soften the blow of the side effects of bioenergy, such as food insecurity, deforestation, and endangered biodiversity; the timing and size of subsidies for renewables; the socially acceptable location of wind-power turbines; the benefits and the risks of nuclear energy; the appropriate volume of international technology transfer; the potential and risks of negative emission technologies (carbon dioxide removal) or solar radiation management; and the employment effects of a rapid diffusion of electric cars.

Several ineffective and inefficient climate policy decisions have already been made precisely because such questions were not answered well. The complexity of these questions requires serious, integrated assessments to facilitate a learning process about the available policy pathways. This presupposes collaborative, scientific explorations of the various societally relevant ramifications of different pathways from different disciplinary perspectives and viewpoints.[9] A more complex version of Pascal's Wager on climate policy is needed particularly when assuming the irreversibility of some climate damages. The burden of proof in risk societies thus mainly lies on assessments of policies to show that a particular climate policy choice is better than its viable alternatives. Such an assessment has to consider a policy's overall effects, side effects, and co-benefits that must be gauged against the backdrop of Sustainable Development Goals and other policy goals or values. Jointly with the stakeholders and decision-makers involved, scientific experts would act as cartographers of policy alternatives while decision-makers remain the navigators.[10] We highly appreciate Oreskes's thoughts concerning the trustworthiness of (climate) science. Yet we stress the need for more focus on the policy assessments to facilitate a better public debate about different solutions and their implications.

The frequently contested value judgments implied in ex-post and ex-ante evaluation of policies are certainly one of the key challenges in determining their trustworthiness and legitimacy. As Oreskes mentions, facts and values—including cognitive, epistemic, and ethical values—are always intertwined in scientific research.[11] It is somewhat surprising that Oreskes does not discuss at greater length how objectivity can emerge concerning disputable ethical values implied in scientific knowledge. Oreskes only points to her hope that there is perhaps much more overlap between our deepest values than we often think—a rather remarkable belief in ethical common ground. Talking about our values and identifying overlap is indeed desirable. She admits, however, that substantial dissent remains regarding some fundamental values and, particularly, their more specific meaning with respect to current policy choices that usually exhibit complexity and uncertainty. Divergent sets of values also play a central role concerning the increasing political divide in several Western countries.

There is good reason to believe that rational discussions about value-laden policy issues are possible, even if they have become partly ideological and almost "religious" disputes. Centuries ago, Pascal made the revolutionary attempt to initiate a rational discussion about nothing less than the most fundamental religious question, namely the very existence of God. If it transpired that God did not exist, the personal costs and benefits of having lived a religious life are comparatively lower than not believing in God, Pascal reasoned. Despite several shortcomings of Pascal's pragmatic wager, including the narrow set of alternatives presented, the latter was one of the first conceptual frameworks for decision-making under uncertainty. John Dewey's (1859–1952) pragmatist philosophy helped to develop this framework further.[12] Similar to Pascal, Dewey emphasizes the philosophical necessity of exploring and evaluating the diverse practical consequences of particular hypotheses, including normative, methodical, and empirical assumptions. All scientific claims and other hypotheses are conceptualized as means of achieving practical ends that are somehow relevant to humankind, assuming an ends-means continuum.[13]

Dewey argues for the possibility of trustworthiness and objectivity of value-laden but always fallible scientific claims— including those of assessments of policies or ethical debates.

On the basis of Dewey's viewpoint, hypotheses—as potential means for resolving problematic situations—can be regarded as trustworthy if they turn out to be reliable in terms of their practical consequences, i.e., if they repeatedly help to transform an indeterminate problematic situation into a determined one in a reliable way. According to this "natural realism," successful results of a pragmatist inquiry could be applied to similar situations; indeed they can serve as the premises for further inquiries as cumulative experience, potentially qualifying as objective, "warranted assertability" under good-enough conditions for such an inquiry.[14] Ends cannot justify the means and both have to be critically assessed via their practical consequences. If, for instance, even the best available climate policy has severe side effects, then the initial policy goals or even the underlying values might have to be revised.

This Deweyan perspective on trustworthiness, albeit largely in alignment with Oreskes's view, offers a nuanced amendment. While the idea of experimentation is inspired by the success of natural science it is applied to all sorts of inquiries—including highly value-laden ones. Moreover, the decisive "practical implications" go well beyond instrumental ones and rather include *everything* that matters for human existence, for instance in spiritual terms.

Instead of presenting allegedly value-free facts or lobbying for a particular policy option scientists can help facilitate more constructive discussions among stakeholders about highly value-laden policy issues like climate policy. Drawing on Dewey's thoughts we propose to go further than normative transparency in scientific assessments and consistently embed divergent values and principles like equality, liberty, purity, nationalism, etc. in different future scenarios and policy pathways. In an inter- and transdisciplinary manner jointly with stakeholders, the various practical implications of these alternative policy pathways can then be critically compared and evaluated.[15] These implications include costs and benefits in the narrow economic sense. Beyond that they also include everything that matters for a society and threatens the legitimacy of policies. "Costs" can, for instance, be perceived as prohibitively high when fundamental rights or procedures are violated.

Such an evaluation of policy pathways may lead to revisions or, at least, reinterpretations of initially assumed, perhaps one-sided sets of values, principles, policy goals, etc. This could happen if it turned out that there were adverse side effects and considerable limitations, e.g., by neglecting other societally relevant values. This exercise can thus enable diverse stakeholders to clarify their policy positions. It can also facilitate the identification of previously unidentified overlap between different viewpoints at least on particular policy instruments and pathways. Left-wing liberals and right-wing conservatives, for example, may still agree about effective carbon pricing.

The essence of the Deweyan approach to value-laden policy assessment consequently lies in its ability to transform heated, entrenched policy conflicts into much more constructive discussions and learning processes about policy alternatives and their complex practical implications. Epistemic trustworthiness can be achieved through a careful exploration and evaluation of the diverse direct and indirect implications of future policy pathways—i.e., of hypotheses conceived of as means within an ends-means continuum—via the feedback loop between ends, means, and consequences. Moreover, the legitimacy of these assessments can mainly be fostered by exploring alternative future policy pathways, which represent different prominent sets of values and policy beliefs, and by actively engaging with a diversity of stakeholders during the assessment process.

The proposed model is largely different from both current practices and the literature regarding the science-policy interface. For instance, our model emphasizes that there is no "value-neutral way" to assess policy pathways and that the feedback loop between policy goals, means, and their practical implications is crucial. Usually, only small sets of alternatives are explored based on narrow evaluation criteria without serious transdisciplinary collaboration or explicit assessing of the underlying policy goals and ethical values via their practical implications. Instead of mere discussion or brokerage of alternative options, the cartography of the implications of alternative futures puts emphasis on learning among all actors involved. This includes learning about alternative problem framings and worldviews.

To conclude, in public policy processes concerning wicked problems and large-scale risks such as climate change, trust in integrated policy assessment is central for decision-making under uncertainty. As an amendment to, and specification of, Oreskes's compelling arguments concerning trust in scientific expertise, we argue that even these assessments of policies can be trustworthy and legitimate despite the controversial value judgments involved. An interdisciplinary and multi-stakeholder exploration of alternative future policy pathways and their various practical implications is required to facilitate legitimate learning processes about the pros and cons of specific pathways. In the end, this could lead to revisions of initially fixed values, policy goals, and means and to the identification of areas of practical overlap between divergent sets of values. Instead, merely insisting on scientific "facts" or criticizing right-wing policy beliefs and values on an abstract level leads to fruitless ideological controversies. We rather need collaborative and inclusive learning processes about alternative futures that acknowledge and critically explore a range of values—as a promising response to the renaissance of populism that is so much based on divergent sets of values.

Chapter 6

COMMENTS ON THE PRESENT AND FUTURE OF SCIENCE, INSPIRED BY NAOMI ORESKES

Jon A. Krosnick

Inspired by Dr. Oreskes, my comments in this essay come from the perspective not of a historian or an expert on the philosophy of science but instead from the perspective of a practitioner of science, observing the present and future of our enterprise.

I believe in the scientific quest for truth, and I believe in the scientific method. I'm glad that contemporary societies value scientific investigation, fund our work, and give us prominence in the news media. I want more young people to choose careers in science. I want scientific disciplines and professional associations to thrive. I want funding of science to increase. And I am looking forward to seeing how scientific discoveries unfold in the coming decades in constructive ways.

To help science to thrive, I cofounded, with Professor Lee Jussim from Rutgers University, the Group on Best Practices in Science (BPS) at the Center for Advanced Study at Stanford University. There are countless stories of scientific successes over many years, and in numerous instances science has gotten off track for a little while before getting back under control eventually. So one can look glowingly at the long-term history of science and smile. But very recent history tells a more distressing and alarming story. And the problem now is not a particular finding that is wrong. During the last decade, we have discovered numerous inefficiencies in science across many disciplines, and dramatic reform is needed, as I will outline in this essay.

My story begins in the field of my PhD, social psychology, with Diederik Stapel, who was the focus of a story in the *New York Times* because he had fabricated the data in more than one hundred publications in top journals in psychology.[1] After this was discovered, numerous papers were retracted, and young coauthors suffered significantly in the process.

Daryl Bem, a very well-known social psychologist at Cornell University, published a paper in the top journal of social psychology claiming to show that extrasensory perception, ESP, was real.[2] It set off a firestorm, because the results seemed implausible from the start and could not be reproduced.

John Bargh, a professor at Yale University, produced a huge amount of beloved work in social psychology. When a group of young scholars sought to reproduce a finding and failed, this led to widespread concern about the replicability of other findings as well.[3] Daniel Kahneman, who won the Nobel Prize in economics, urged Dr. Bargh to engage with the critics of his work to pull the field toward an understanding of which empirical findings are real. But no such reconciliation has yet occurred.

At one time the most downloaded article in the history of the *New Yorker* magazine was an essay written by Jonah Lehrer about the so-called decline effect.[4] In the piece, psychologist Jonathan Schooler explained how he discovered an important phenomenon, called verbal overshadowing, but the more he studied it, the weaker and weaker the effect became, until it disappeared entirely.

Another landmark paper described what were called "voo-doo correlations."[5] In studies of neuroscience, gigantic correlations were being published between brain functioning and other indicators of psychological experience, and those correlations later appeared to have been the fabricated results of manipulative research practices.[6]

Consider Phil Zimbardo's prison experiment, in which a group of participants were randomly assigned to be guards or prisoners in the basement of the psychology building at Stanford.[7] The BBC, a few years ago, attempted to conduct the same study and failed to observe the same results.[8]

One of the most famous studies in social psychology, by Leon Festinger on cognitive dissonance, documented the difference between how people reacted to a task when they were paid one dollar versus when they were paid twenty dollars.[9] This study has been cited numerous times over the years but has never been successfully replicated, as far as I know. Most importantly, Festinger himself is famous for having said that he had to run the study numerous times, tweaking the methodology, before he got it to "work"—that is, to produce the result he wanted.

Yet another example: a paper on what's called biased assimilation and attitude polarization.[10] The authors concluded that if a person reads a balanced set of evidence, with about half supporting a particular conclusion and the other half refuting that conclusion, the person evaluates the evidence so as to protect his or her predilections. As a result, reading a balanced set of evidence was said to have made people's opinions on the matter even more extreme than were their original views. But A. G. Miller and coauthors showed that the original paper was incorrect because it had used an improper measurement approach.[11] The original paper has been cited more than 3,000 times, and Dr. Miller's paper has been cited just 136 times. This is not an example of science correcting itself successfully.

These are not isolated incidents, cherry-picked to tell a pessimistic story. Consider a story in the *New York Times* under the online headline "Many Psychology Findings Not as Strong as Claimed, Study Says."[12] This study, conducted in 2015 by Brian Nosek and colleagues, attempted to replicate the findings of many highly prestigious, randomly selected publications in psychology. When the story appeared in print, the headline was "Most Psychology Findings Cannot Be Replicated." And that is, in fact, the conclusion of the paper. A random selection of studies, an aggressive effort to try to reproduce the findings, and majority failure.

This problem is not confined to my discipline of psychology alone. Consider political science. A paper entitled "When Contact Changes Minds" was published in *Science* exploring the idea that conversations on doorsteps could change attitudes about gay marriage.[13] The paper was also written about in the *New York Times* following the discovery that although the principal author claimed to have collected data, he had not.[14] The whole study had been fabricated. In economics, a series of papers show that in attempts to replicate the findings of empirical studies, the majority could not be reproduced.[15] And surveys of voters failed spectacularly in recent efforts to predict the outcomes of elections in the United States, Britain, and Israel.[16]

This problem has been illustrated in the physical sciences as well, especially vividly by Amgen Pharmaceuticals, whose scientists attempted to replicate fifty-three landmark findings, published in the most prestigious journals, *Science, Nature,* and *Cell.*[17] Amgen had tried to develop new drugs based on such findings and had failed repeatedly. So Amgen stepped back to the fundamentals—to determine whether they could trust what they read in these journals. A team of one hundred scientists found that 89 percent of the findings they tried to reproduce could not be reproduced.[18] When the Amgen scientists told one study's author that Amgen had tried numerous times to replicate his or her finding and couldn't, the original author said that he or she had failed a good number of times before finally producing the desired findings.

When Amgen went public, Bayer Pharmaceuticals reported having had the same experience.[19] They tried to replicate sixty-seven published findings, and 79 percent of them could not be reproduced. And problems have become vivid in chemistry as well. In recent years, an increasing number of publications included doctored graphs that made findings look better than they were.[20] And in ecology, genetics, and evolutionary biology, findings were published and then disappeared, never to be reproduced.[21] In line with all of these trends, the website called Retraction Watch has documented a skyrocketing number of retractions of published articles.

When the BPS group at Stanford spoke to engineers about their experiences in this regard, we were paralyzed by a shocking comment. When asked whether there are problems with reproducibility and integrity in engineering, they replied, "Truthfully, we don't believe the findings of any other labs." So we asked, "Do you believe the findings from your lab?" and they said, "Sometimes."

Why? In engineering, it's not uncommon for authors to intentionally leave out a key ingredient of the formula needed to make the soup, so that the competing labs can't get ahead of them in the race to build, for example, a battery that lasts longer.

What about in a field where lives are at stake, medicine? John Ioannidis has been meta-analyzing research findings in health research for decades. One of his papers gauged the reproducibility of preclinical medical research and found that more than 50 percent of the studies could not be replicated even once. This lack of replication wastes $28 billion each year in research that leads to nothing.[22] This publication, "Why Most Published Research Findings Are False," was at one time the most downloaded paper from the journal *PLOS One*.[23]

Why is all this happening? How can it be that science is wonderful when all of these examples pile up to suggest we're in trouble? One answer is that contemporary scientists engage in a variety of counterproductive, destructive practices.

One is called p-hacking: manipulating and massaging data in order to get desired results. Another problem is reliance on small sample sizes. If a study with a small sample doesn't work out, an investigator can discard it and do another small study until desired results are obtained by chance alone, because the cost of doing each study is low. Another problem is improper calculation of statistics, leading to overconfidence in the replicability of a finding. Some disciplines have been in the habit of computing statistical tests in ways that are knowingly biased toward getting significant findings. If computed properly, the statistics would suggest more caution. And in some physical science fields, experiments never involve random assignment to conditions and do not involve statistical significance testing, which leaves investigators vulnerable to being misled.

Of course, accidental mistakes in statistical analysis will occur, so proofing is essential. But when a scientist has obtained desired results, it's tempting to celebrate with optimism. And if undesired results are obtained, perhaps scientists are more motivated to check their work and more likely to detect errors. Thus perhaps errors go uncaught when results are desired.

How prevalent are suboptimal practices by scientists? In one survey, a large number of psychologists said that they have implemented many suboptimal practices.[24] So perhaps the prevalence of irreproducible results should be no surprise.

All of us in science confront this reality by marshaling a desire to remedy the problems. And in order to do so, we need to know what causes suboptimal behavior on the part of scientists. And, unfortunately, the causes are horrifying. There are individual-level motivators of scientists themselves. Many scientists want to be famous, to get research grants, to be employed with tenure, to get outside job offers to increase their salaries, to get promoted, to be well paid, to found mega-profitable start-ups, to be respected by their peers, to be respected by non-scientists, and more.

If you put a scientist in a quiet dark room by himself or herself, he or she will most likely acknowledge that, almost always, we operate in an environment (in academia or outside) where these motivations are very clearly powerful for everyone. So we don't need to know whether a project's funding comes from ExxonMobil or the National Science Foundation. We're all operating in an environment in which we have these motivations.

Everything's fine if these motivations are coupled with a desire always to be right, always to publish the truth. But unfortunately, system-level causes carry us off that path. Systems value productivity, reward faculty who publish a lot, and don't reward faculty who publish less. And most disciplines value innovation and unexpected findings. At Stanford, graduate students in psychology have been taught not to waste their time publishing something that people already thought was probably true. The goal should be to produce findings that would surprise people. And if that's the goal, should we always be asking ourselves whether a really surprising finding is surprising because it contradicts existing theory and evidence, and is unlikely to be real?

Systems value statistically significant findings, so journals have been biased against publishing null findings, which slows down the process of discovering that previously published findings are false.

Researchers want to publish a lot: they want to publish novel, counterintuitive findings; they want to tell a good story; they want to defend their own previous publications and reputation; they don't want to admit that they didn't predict their findings and that they were surprised. They want to disseminate findings as quickly as possible, and they attract the attention of the news media.

Institutions encourage all this, because universities are increasingly using metrics counting publications and citation accounts in tenure and promotion decisions. Journals are biased against messy results, where study findings are inconsistent from one to the next. Journals disseminate innovative findings more quickly. News reporters sometimes cause problems as well, by asking for a simple, general conclusion when a qualified claim would be more justified. Journals have page limits, despite the fact that we no longer need paper to disseminate our work. Page limits restrict the degree to which we can be transparent about our methodology. And journals are favorable to some types of findings, in some cases, that favor particular political agendas. And, of course, research assistants often want to make the PIs happy, which can create motivations to produce certain findings instead of others.

My observations above are mostly speculations. Everything I have said could be wrong. But I might be right as well. As far as I know, no one is testing a theory like this to explain the behavior of scientists. And we need that sort of testing, because we now know that scientific literatures and popular media are filled with findings that cannot be reproduced. We don't know as much as we say we know about the matters that we study. And if that's true, we need to embrace the problem, and we need to get to work on solving it by implementing reforms that empirical evidence shows will work.

So this is a problem of social and behavioral science. It's a problem of human psychology. It's not a problem of chemistry. It's not a problem of physics. It's not a problem of intuition. And it's a problem that requires empirical research, informed by theory and using rigorous methods of investigation.

What about solutions? Many have been proposed, but in my opinion they are mostly band-aids. They make people feel good in the short term. But we don't yet know whether they actually work to increase the efficiency of scientific investigation.

What conclusions can we reach? First of all, I don't think that the source of funding is a primary problem for science. In fact, I would guess that's among the least of our problems. A huge amount of research has been funded by federal agencies and private foundations that have no real agendas other than supporting scientists' making discoveries as quickly as possible.

Rather than funding sources, the fundamental problem is with the incentives inherent in the world in which science operates today. Blanket discounting or acceptance of findings based on who paid for them is probably missing the point. The problem is that new technology has sped up the process of science. We hoped that technology would make science more efficient. But instead, science is either operating incredibly inefficiently or publishing a vast majority of findings that are false.

What is the path forward? First, we must acknowledge the problem. It does a patient no good for the doctor to withhold the information that he or she has cancer. Second, we have to identify the real causes of problematic behaviors, instead of speculating. Third, we need to develop solutions to undermine the counterproductive motivations that drive science in the wrong directions. Last, we should be scientists, and we should test the effectiveness of those solutions.

I hope that these thoughts, inspired by Dr. Oreskes's lectures and essays, complement her essays with a focus on more contemporary history and the present of science. In offering these thoughts, I hope to encourage all scientists to consider this to be a good moment to stop and reflect, to try to learn from the past of science, and to redirect the present and future of science in ways that considerably enhance the efficiency of the enterprise in achieving its goals.

RESPONSE

Chapter 7

REPLY

Susan Lindee has given us a brilliant exposition of the ways in which late twentieth-century scientists built a narrative stressing the distinction between science and technology and the reasons why they did so. During the Cold War, scientists' ambivalence toward the project of building a massive nuclear arsenal expressed itself (in part) through their insistence that science and technology were separate domains. Noncompeting magisteria, we might say, to borrow Steven Jay Gould's famous formulation of the relation between science and religion, except that whereas Gould urged respect for both domains, scientists, Lindee argues, were constructing a thesis of separate and unequal: science was separate and superior to technology because of its moral purity. To retain that purity, it needed to stay separate.

Historians of science recapitulated this framework, denying the tight linkages and intellectual affinities between science and technology and emphasizing instead their distinctive characteristics, independent institutional structures, and mostly nonoverlapping populations or practitioners. Historians of technology rejected the premise that their object of study was inferior, but accepted—indeed promoted—the notion that it was separate and distinct from science. Both groups were content to see their professions proceed on parallel tracks, and many preferred it.

I think Professor Lindee is completely correct about this. In my own recently completed magnum opus on the history of US

Cold War oceanography, I have made a similar argument: that American oceanographers generally downplayed and sometimes denied the technological aspects of their work.[1] Most of these technological elements were closely related to submarine warfare, including the delivery of weapons of mass destruction (to use our language).

For the most part scientists were unable to discuss these relations because of security restrictions, so it is not always easy to discern how they felt about them, but some scientists explicitly expressed moral qualms. Others did not necessarily doubt the imperative of countering the Soviet threat, but nevertheless questioned the wisdom of hitching their scientific horses to the military wagon. One way to skirt the moral dimension was to insist that their work was not so hitched: that even though the US military was paying for it, scientists had retained control of their intellectual agenda.[2] The ideological framework of pure science enabled them to claim—and perhaps believe—that the knowledge they produced was separate and distinct from anything the US government, through its armed services, might do with it. And so many of these oceanographers insisted they were pursuing "pure science," even when this was manifestly not the case.

Given this history, it is not surprising that many Americans do not have a clear conception of the historical or current relationships between science and technology. But I suspect that our current situation is overdetermined: there may be many reasons why Americans are confused about science and technology. These include a near-complete lack of engineering education in primary and secondary schools so that most students, unless they study engineering in college, will have no exposure whatsoever to engineering and no sense of how engineers use science in their everyday work. Conversely, science is generally taught with little reference to its practical uses, and popular science writing perpetuates a myriad of myths that distress historians, the relation between science and technology being from my perspective the least of them.

Thus, it seems to me unlikely that what scientists or historians claimed in the 1950s and '60s is a primary factor explaining our current situation. To be sure, today's senior scientists were raised by the Cold War generation and perhaps for this reason have often perpetuated the "separate and unequal" framework that Lindee laments. But the current generation has also pioneered biotechnology—which by its very name declares itself to be both science and technology—and routinely invokes technology as a justification for why we should believe in science.[3] Yet this has not stopped religious fundamentalists from rejecting evolutionary theory nor free market fundamentalists from rejecting the facts of anthropogenic climate change.

Yes, there is a tremendous amount of scientific knowledge embedded into everyday technologies, from roads and bridges to iPhones and laptops, and, indeed, frozen peas. Explaining this more clearly in public settings and in the classroom would remind people that we have direct evidence of science in action in everyday life. But I doubt that it would have the effect Lindee thinks, because even if people are well informed of the science embedded in their cell phones, it is not likely that will change their stance on climate change.

The reason for this is clear and well established: Americans do not reject science, *tout court*, they reject particular scientific claims and conclusions that clash with their economic interests or cherished beliefs. Numerous studies have shown this to be true. The recent report of the American Academy of Arts and Sciences, for example, showed that most Americans do not reject science, overall, but do reject evolutionary biology if they interpret it as clashing with their religious views, or climate change if they see it as clashing with their political-economic views. Tellingly, many Americans are quite content to accept that DNA carries hereditary material even while rejecting evolution as a process that over time alters the DNA of populations.[4]

Moreover, this pattern is not a uniquely American pathology, nor a particular feature of the present moment. In the twentieth century, Einstein's path-breaking work on relativity was rejected by Germans who felt it threatened their idealist ontology.[5] Vaccine resistance has been going on for just about as long as vaccination has. Smallpox vaccine noncompliance was so widespread in late nineteenth-century England that the Vaccination Act of 1898 included a "conscience clause" allowing parents to decline vaccination on grounds of personal belief.[6]

If we explain to people how their cell phones work, they may very well feel better about those phones, but absent other interventions it will likely have little or no impact on their views of evolutionary theory. People compartmentalize.

Perhaps most fatally to Lindee's argument, we know that in recent decades various parties have cynically exploited the values that lead some Americans to reject climate science or evolutionary theory for their own social, political, or financial ends. Recent work suggests that people's opinion and attitudes can be shifted if you show them—with concrete examples—that this is the case, and explain to them how disinformation works. John Cook and his colleagues call this "inoculation": By analogy with vaccinations, if you expose people to a small amount of disinformation, in a controlled environment, you can generate future resistance.[7]

Lindee's proposed solution also conflates utility with truth. This is a point that philosophers have long emphasized. When we say that something works, we are making a claim about performance in the world. Cell phones enable us to talk to people in other locations without being connected by wires. Laptops enable us to store and access huge amounts of information in a very easy way. Vaccinations prevent diseases. Frozen vegetables enable us to eat foods long after they were harvested. No one doubts that these things do the things they claim to do, because we see that it is so. But it is another thing to claim that this proves the truth of the theories that underlie them.

We might argue that every time we use a piece of technology we are performing a small but significant experiment, confirming that the technology works. (Or not, as the case may be. Frozen peas taste pretty lousy.) But this is a very different matter from confirming the underlying theory required to design, build, and use that technology. There are many reasons for this: suffice it here to consider three.

The first is that that my cell phone does not reify a single theory. My phone is a complex expression of the many diverse scientific theories and practices that have gone into its development. These could include theories from electromagnetics, information technology, computer science, material science, cognitive psychology, and more, as well as various engineering and design practices. We might argue that the success of the cell phone affirms the correctness of all these theories and practices, but the conceptual link for the average user will be vague at best.

The second is that Lindee's premise—that the success of the technology is proof of the theory behind it—is an instance of the fallacy of affirming the consequent. It is the core of the logical objection to the hypothetico-deductive model that we discussed in chapter 1. (To recapitulate: If I test a theory and it passes my test, this does not prove that the theory is correct. Other theories may have predicted the same result. Or two [or more] errors in my experiment may have canceled each other out. It is a fallacy to assume that, because my theory worked in that instance, I have demonstrated it is true.)

David Bloor has provided a third powerful argument against Lindee's line of reasoning, which we might label the fallacy of theoretical precedence. Consider airplane flight.

We might think of airplane flight as one of the most obvious examples of the success of science. For centuries (perhaps longer), people dreamed of flying. Birds could fly, and so could insects. Some mammals could glide over considerable distances. Why not us? In the early twentieth century, clever inventors overcame the challenge of heavier than air flight, and soon we had commercial aviation. Today airplane travel is as familiar to most Americans as frozen peas. It is just the sort of everyday technology in which Lindee places epistemological aspiration. She is not alone. In the 1990s, when some scientists tried to defend realist concepts of scientific theory in the face of the challenge of social constructivism, airplane travel was a favorite invocation. How could planes possibly fly if the theory behind them was just a social construction? If it were something that scientists had settled upon for social rather than empirical reasons? If it wasn't *true*?

Confronting this argument, David Bloor uncovered a startling historical fact: that engineers were building planes before they had a working theory of flight. In fact, heavier-than-air machines were flying for years while existing aeronautical theory declared it impossible. As Bloor explains: "The practical success of the pioneer aviators still left unanswered the question of how a wing generated the lift forces that were necessary for flight." The technological success of aircrafts did not signal an accurate theoretical understanding of aerodynamics.

Perhaps the history of aviation is an odd exception, where technology got ahead of theory. But one of the contributions of the history of technology during the period when it was severed from the history of science was to show how many technologies, particularly prior to the twentieth century, developed relatively independently from theoretical science. Many technological innovations were empirical accomplishments whose relationship to "science" was only established retrospectively.[8] Following Lindee, one might argue that one can nonetheless use the success of technology to build trust in current relevant science, irrespective of historically contingent relationships. Or one might suggest that what held true in the eighteenth and nineteenth centuries, and perhaps even in the early twentieth, is no longer the case, and it is simply implausible that our exquisitely complex modern technologies could work as they do if the theories behind them were not true. Perhaps this is the case. Only time will tell.

Marc Lange suggests that the question of why we should trust science can "easily induce a kind of dizziness or even despair." Many potential answers collapse into circularity. For example, reasoning that invokes empirical evidence (such as my argument based on history) is itself a form of scientific (i.e., empirical) argument, in which case we are using scientific styles of reasoning to defend scientific styles of reasoning—the very definition of circularity. Moreover, if I say we should trust science as the warranted conclusions of experts, then we must ask on what basis is someone to be judged an expert? The answer, of course, is by other experts. So that is circular, too.

Or is it? The signs of expertise—academic credentials, publications on the pertinent topic in peer-reviewed journals, awards and prizes—are evident to non-experts. Journalists have sometimes asked me, "How am I to tell if an alleged expert really is one, and not just a shill?" I reply, "One place to start is to find out what field they trained in and what publications they have in the domain." Of course, Professor Lange is right to note that training is provided by other experts—it takes an expert to make an expert—and so it may appear that we have not escaped circularity. But there is an escape, because the social markers of expertise are evident to non-experts. This is a non-trivial point, because it is relatively easy to discern that most climate change deniers are not climate scientists, and that objections to evolutionary theory largely emerge from non-scientific domains. Neutral non-experts can identify experts and discern what they have (or have not) concluded.

Social markers do not tell is if an expert is trustworthy, but they do tell us if the person is an expert and, more to the point, if a person *claiming* expertise does not possess it. Similarly, it is (or should be) easy to distinguish a research institution—like Princeton University or the Lawrence Livermore National Laboratory—from policy-driven think tanks, such as the American Enterprise or Discovery Institutes. The fact that journalists often fail to make such distinctions has more to do with deadlines than with epistemology.

Of course, as I have stressed, experts can be wrong. (Our entire inquiry would be superfluous were this not the case!) As a human activity, science is fallible. Consensus is not the same as truth. Consensus is a social condition, not an epistemic one, but we use consensus as a proxy because we have no way to know, for sure, what the truth is.

Moreover, the category of consensus is epistemically pertinent, because our historical cases have shown that where experts appear to have gone astray, typically there was a *lack* of consensus. Thus, we need to live with the fact that our indicators are asymmetric. We can never be absolutely positively sure that we are right, but we do have indicators that suggest when something might be wrong.

This is why consensus is important—why it is so important to be able to identify and discount shills, celebrities, and perhaps well-meaning but misguided lay people—in order to clarify who is an expert, what they have to say, and on what basis they are saying it.

Lange's own solution to these dilemmas is to examine how past debates were resolved and agreement achieved. He shows us that even in the heat of debate, it can be possible to find an argument that persuades one's scientific agonists. Galileo persuaded contemporaries of the superiority of his proposal for the relation between time and the distance traveled by a moving body by showing that only his proposal satisfied the demand of dimensional homogeneity, i.e., being independent of the particular (and presumably arbitrary) units applied. Lange concludes that "finding powerful reasons in a crisis is inevitably going to be difficult, but not impossible."

This argument, drawn from history, is a nice one, but (as Lange himself acknowledges) it still leaves us with the problem of generalizing from specific examples to science as a whole. Perhaps there is no way to so generalize, but Lange agrees that we ought to try. And that, of course, is the point of this book.

Ottmar Edenhofer and Martin Kowarsch address the relationship between scientific knowledge and public policy. In a world where dangerous anthropogenic climate change threatens human life, liberty, and property, the future of biodiversity, and the stability of liberal democracy, the relation between science and policy is of no small concern. Yet, the scientific consensus on climate change has not led to policy consensus. Indeed, they suggest that it cannot, because policy decisions involve many more dimensions than scientific findings do. In particular, policy decisions entail value choices above and beyond whatever values may have been embedded in the scientific work. Thus, they conclude, more work is needed on the "role of scientific expertise and the design of policies" to understand how we get from science to policy on urgent, contested, value-rich issues.

All this is true, but a bit orthogonal to the point of this book.

I posed the question—*Why Trust Science?*—because in recent decades, some groups and individuals have actively sought to undermine public trust in science as a means to avoid policy action that may be warranted by that science. This includes but is by no means limited to climate change. In the United States, it includes such diverse matters as the justification for compulsory vaccination, the hazards of persistent pesticides, and whether children raised by homosexual parents turn out as well adjusted as those raised by heterosexual ones (or at least no less well-adjusted). At least in the United States, where I have studied the matter most closely, it is confusing to say, as Dr. Edenhofer and Kowarsch do, that "conflicts about climate policy are thus not necessarily rooted in a lack of trust in climate science—but rather in disagreement about the design of climate policies." My work has shown that (for the most part) they are not *rooted* in a lack of trust in science at all! They are rooted in economic self-interest and ideological commitments, and are intended to stymie discussion of climate policies.

As Erik Conway and I showed in our 2010 book, *Merchants of Doubt*, those who deny the findings of climate science do not (for the most part) have principled disagreements with scientists, economists, and environmentalists about the *best policy* to address anthropogenic climate change. Rather, *they do not want any policy at all.*[9] Because of economic self-interest, ideological commitments to laissez-faire economics, or both, they do not wish to see any government action to limit the use or raise the price of the fossil fuels whose use drives climate change. What they want is preservation of the *status quo ante.* Recognizing that an honest accounting of the costs of climate change would almost certainly warrant a change in the status quo, they attempt to undermine public confidence in the science that supports such an accounting. Discrediting science is a political strategy. Lack of public trust in science is the (intended) consequence.

Given this, it would be absurd for me to expect that articulating the reasons for trust in science would alter the positions of climate change deniers. What is not absurd is the hope that for some readers, this book will answer legitimate questions, even if those questions have at times been raised by people with a political agenda with which I strongly disagree.

Do we need more information on how science is used (or not) in policy? Absolutely. For all the reasons that Drs. Edenhofer and Kowarsch articulate, science alone cannot not tell us what, if anything, to do about disruptive climate change (or any other complex social challenge). But the natural sciences do tell us that if we continue business as usual, sea levels will rise, biodiversity will be lost, and people will get hurt. The social sciences further tell us that trillions of dollars will be spent dealing with climate damages—money that could be used in happier and more productive ways—and we are all going to end up a lot poorer. The point of this book is to explain how and why this science is likely to be worthy of our trust. Should we go further and think harder about the basis for trust in the scientific assessments, such as the reports of the Intergovernmental Panel on Climate

Change, which attempt to collate, judge, adjudicate, and otherwise evaluate scientific evidence for the purpose of informing policy? By all means.

The fact that I did not address this issue in these chapters should not be taken as disparaging its significance. On the contrary, I have written a different book on that topic![10] But since it is bad form to answer a serious question by saying "read my other book," let me say a few words here.

Assessment for policy occurs in a different context from everyday science: namely, that a problem has been identified and some agency of governance has asked for information to help guide policy choices. Often this involves deadlines. Reports are needed in time for a particular press conference, congressional session, international meeting, or the like. There is a demand for an answer, even if the science required to supply it may be evolving and incomplete. This—along with their complex moral and political landscapes—makes assessments for policy more complex than science left to its own devices.

This is not to say that everyday science does not also encounter real, suspected, or alleged problems and threats: Edward Clarke certainly believed that female higher education was a threat. But there is a major difference: the IPCC was formulated as part of the UN Framework Convention on Climate Change, which formally recognized anthropogenic climate change as a threat to sustainable development. (In this sense, a value premise—the value of sustainable development—was embedded into its creation.) This instrument of international governance asked the scientific community, qua community, to give its best assessment of the consensus of scientific opinion—the state of scientific knowledge—relative to this challenge. No one asked Dr. Clarke for his opinion on female scientific education; no institutional body was waiting for his answer. Thus, one obvious response to Clarke's work is to point out that it was a single study by a single author; there was nothing remotely approaching a consensus on the matter at hand: not on the proposed solution and not even on the existence of the alleged problem.

I endorse Edenhofer and Kowarsch's view that the complexity of climate policy issues requires "serious, integrated assessments to facilitate a learning process about the available policy pathways," and that any such process must necessarily include both the natural and social sciences, as well as perspectives from law, government, religion, and the humanities. Their position is entirely compatible with the arguments I have presented. I also agree that values cannot be excised from such discussions; value differences are a central reason why we have political and social conflict. But I do maintain that ethical overlap among agonists is often greater than may appear. Edenhofer and Kowarsch implicitly acknowledge as much when they invoke "fundamental rights" and the (implicitly undisputed) "backdrop of Sustainable Development Goals," as well as the prospects for "rational discussion about value-laden policy." We don't all agree on everything, but many of us agree on some things, and some of us agree on many things.

My argument complements theirs, insofar as we are all arguing for the open discussion of the role of values in both science and policy. I am not naively suggesting that if only we are transparent about values, all will be right with the world. I am arguing that if we can make the overlap in our values explicit, this may, in some cases, help us to overcome distrust that is rooted in a *perceived* clash of values. But this is hard for most natural scientists to do.

Because they have been enculturated in the norms of value neutrality, most scientists feel the need to hide or expunge their values from scientific practice and discussion.[11] I argue that this is unnecessary and possibly counterproductive. For example: many scientists, even if they are not themselves religious believers, hold values that overlap with believers. This was demonstrated in a series of meetings at the Pontifical Academy of Sciences, which helped to lay the foundation for the Papal Encyclical on Climate Change and Inequality, *Laudato Sí*.[12] The scientists and theologians who attended those meetings did not by any means agree on all things, but we found considerable common ground, which Pope Francis made explicit in his writings.

Diverse people will never agree on all things. My Christian friends believe in the divinity of Jesus Christ and I do not. That is unlikely to change. My point is not that we will reach theological or ethical consensus, but only that, if we share some values, then we can find common ground for a conversation. And that may help us to overcome what otherwise appears to be an insurmountable divide, not only on climate change, but perhaps on other matters as well.

Professor Jon Krosnick calls attention to a serious issue in contemporary science: the "replication crisis."[13] It is an issue with potential to undermine public trust in science, as well as to refute my argument that the communal processes of vetting scientific claims are likely to lead to reliable results so long as the vetters are diverse and open to self-criticism.

The issue is this: there have been a number of well-publicized examples of papers published in reputable journals—and in some cases heavily cited—whose results could not be replicated. Some papers have been retracted, leading commentators to declare a "retraction crisis."[14] Much of the discussion of the replication crisis, as well as of potential remedies, has focused specifically on psychology and biomedicine.[15] However, Professor

Krosnick claims that the problem pertains to all contemporary science, because of the incentive structure that rewards rapid publication at the expense of care and diligence. This may be, but Krosnick's specific examples are all from psychology and bio-medicine, and the latter predominantly from clinical trials of drugs. Both are domains in which statistical analysis plays a central role, and both are areas wherein the misuse of statistics— particularly p-hacking—has been demonstrated.

In 2019, a paper published in *Nature* called for a rethinking of the entire manner in which statistical tests of significance are conventionally used in science. They noted that the failure of a test of an effect to achieve statistical significance at the 0.05 level is not proof that the effect does not exist, yet scientists often claim that it is. Likewise, the finding that the difference between two groups does not achieve statistical significance at the 0.05 level does not prove that there is no difference between those groups, but scientists often make that claim too.[16] The authors called for an end to the use of p-values in a dichotomous way and for "the entire concept of statistical significance to be abandoned." Their paper was supported by the signatures of over eight hundred additional scientists, suggesting that these issues are widespread. We might expect that in any field that relies heavily on statistics—particularly where students are taught statistical tools in a "black-box" fashion— these problems might indeed be widespread.[17]

There is evidence of this. In a series of recent papers, my colleagues and I have demonstrated that the misapplication of statistics to historical temperature records—combined with social and political pressures—led many climate scientists to conclude wrongly that global warming had stopped, "paused," or experienced a "hiatus" in the 2000s.[18] Despite our work, the misimpressions persist: One government science agency blog post in 2018 misleadingly posed the question "Why did Earth's surface temperature stop rising in the past decade?" Later, it was updated to inform the reader: "Since this article was last updated, the slowdown in the rate of average global surface warming that took place from 1998–2012 (relative to the preceding 30 years) has unequivocally ended."[19]

This sentence illustrates how scientists tried to save face when it was demonstrated that Earth's surface temperature did not stop rising in the past decade: they altered their terms, replacing stoppage, pause, and hiatus with "slowdown." The latter term reflected the fact that the rate of warming appeared to be lower in the 2000s when compared to a baseline representing the period during which anthropogenic climate change has been underway.

This may seem like a trivial replacement—merely semantics, in some views—but it is not. It is well known that Earth's climate fluctuates, so even in the face of a steady rise in atmospheric greenhouse gas concentrations, the rate at which the planet would warm would vary. No scientist would expect otherwise, but, *ceteris paribus,* we would expect the overall direction of change to remain positive. This is, in fact, what happened. In other words: nothing abnormal or unexpected occurred. The observed slowdown was neither scientifically surprising nor epistemically problematic. It was not something that required explanation. Yet, many scientists treated it as if it were, leading to a great deal of misleading conversation both in the scientific community and in public arenas.[20]

It seems reasonable therefore to conclude that the misuse of statistics is not restricted to psychology and biomedicine. But is there a broader problem with science, writ large? Here the evidence becomes more ambiguous, and I find it surprising that Professor Krosnick—who stresses the importance of rigorous empirical research—makes broad claims on limited evidence and lumps together phenomena that may be distinct.

In his opening, he offers a story of outright fraud—a professor who had fabricated data in over one hundred publications in leading journals. No doubt this is bad stuff, but fraud is a feature of all human activity. Is it more common in science than in finance? Or real estate? Or mineral prospecting? The information offered here does not enable us to judge.[21] What it does enable us to do is to ask why was this fraud not detected sooner, reminding us that science (like every human activity) demands oversight, and to consider whether better oversight mechanisms in science are needed.

Then Krosnick offers us something completely different: the story of a paper that claimed to demonstrate the reality of ESP and "set off a firestorm, because the results seemed implausible and could not be reproduced." This is the opposite of fraud: it illustrates science working as it should. A paper was published that made a strong, surprising, and implausible claim. Immediately it received tough critical scrutiny, and the psychology community rejected it. One might query why this paper was published in the first place, but if science is to be open to diverse ideas (as I have argued it must be), then it is inevitable that incorrect, stupid, and even absurd items will sometimes make their way into print. By itself, that is not an indictment of science. On the contrary, it is evidence that the scientific community has remained open, even to ideas that some of us might think should be closed down.

Then we have the example of one of the most well-known studies in the recent history of psychology—the famous (or infamous) Stanford prison experiment. Here, we are told that the BBC—which is not a scientific organization, so one has

immediately to wonder about motivations and possible bias—
tried and failed to replicate that study.[22] Now we have study 1
versus study 2. What are we to think of that? Four options
present themselves:

> Study 1 is correct and study 2 failed to replicate it because of
> flaws in the latter.
> Study 2 is correct and study 1 should be considered refuted.
> Both studies are incorrect, albeit in different ways.
> Both studies are correct, but the conditions under which
> they were performed were different, and therefore they
> provide different information about the effects of the
> conditions under which humans behave.

Without additional information, it is impossible to determine
which of these four options is the right one.[23]

Most of the studies that Professor Krosnick offers as evidence
of trouble in science are single studies that were later shown to
be faulty. But the thrust of my argument is to stress that scientific knowledge is never created by a single study, no matter how
famous, important, or well-designed. What leads to reliable scientific knowledge is the process by which claims are vetted.
Crucially, that vetting must involve diverse perspectives and the
presentation of evidence collected in diverse ways. This means
that a single paper cannot be the basis for reliable scientific
knowledge. In hindsight we might conclude that the Stanford
prison experiment was given far too much weight, considering
that it was a single study.

Albert Einstein's celebrated 1905 paper on special relativity
is a case in point: many people know only of that paper and
think that Einstein, on his own, overturned Newtonian mechanics. This is an incorrect view, made possible by ignorance
of history. Many of Einstein's contemporaries helped to lay the

groundwork that made the 1905 paper both possible and plausible (most famously, Hendrik Lorenz), and much subsequent work went into consolidating the epistemic gain of the 1905 paper. The same was true of general relativity: various colleagues, including the mathematician Emmy Noether, helped Einstein to resolve difficulties in the theory, and it was the Englishman Sir Arthur Eddington who undertook the experimental confirmation that convinced the world that the theory was true.[24]

Professor Krosnick's commentary thus reinforces my argument about consensus: We should be skeptical of any single paper in science. Scientific discovery is a process, not an event. In that process, many provisional claims—perhaps even most provisional claims—will be shown to be incomplete and sometimes erroneous. As several past presidents of the US National Academy of Sciences recently argued, refutation and retraction, if done in a timely manner, may be viewed as science correcting itself as it should.[25] Conventionally, we have called this process *progress.*

Admittedly, this does put us in a difficult situation when we have to make decisions on the basis of scientific knowledge that may in the future be challenged. What are we to do at any given moment, when we cannot say which of our current claims will be sustained and which will be rejected? This is one of the central questions that I have raised. Because we cannot know which of current claims will be sustained, the best we can do is to consider the weight of scientific evidence, the fulcrum of scientific opinion, and the trajectory of scientific knowledge. This is why consensus matters: If scientists are still debating a matter, then we may well be wise to "wait and see," if conditions permit.[26] If the available empirical evidence is thin, we may want to do more research.

But the uncertainly of future scientific knowledge should not be used as an excuse for delay. As the epidemiologist Sir Austin Bradford Hill famously argued, "All scientific work is incomplete—whether it be observational or experimental. All scientific work is liable to be upset or modified by advancing knowledge. That does not confer upon us a freedom to ignore the knowledge we already have, or to postpone the action that it appears to demand at a given time."[27] At any given moment, it makes sense to make decisions on the information we have, and be prepared to alter our plans if future evidence warrants.[28]

Returning to psychology, I was surprised that Professor Krosnick did not offer what I consider to be the most egregious recent example of bad science in that field: the "critical positivity ratio." This was the claim that a very specific number—2.9013—could be used in a variety of ways to distinguish psychologically healthy individuals from unhealthy ones.[29] After the paper was published in 2005, it was cited over one thousand times before being debunked in 2013, when a graduate student, Nick Brown, collaborated with physicist Alan Sokal and psychologist Harris Friedman on a reanalysis of the data.[30] In hindsight it is bizarre that this paper—with its implausibly broad and ambitious claims and absurdly precise "ratio"—five significant figures!—would have broadly been accepted. Its theoretical reliance on nonlinear dynamics might also have suggested that the paper was little more than trendy hype.[31] But the crucial point here is this: it was a single paper. It may have been heavily cited, but it did not represent the consensus of professional experts.

Perhaps Professor Krosnick did not include it because it does seem to suggest that something is rotten in the state of psychology. But that is not the conclusion Krosnick wants and, perhaps for this reason, he paints with a broad brush and speaks in general terms. I think this is unfortunate, for it does not help us to delineate the extent and character of the problem. He gives us a singular example of fraud in political science and uses this to implicate the entire field. He speaks of the "physical sciences" when he is referring to biomedicine. He suggests that problems in engineering are "not uncommon," but then offers only hearsay and anecdotes about engineering, and no evidence at all from physics, physical chemistry, geology, geophysics, meteorology, or climate science. He acknowledges that his observations about the causes of the alleged crisis are "mostly speculations."

Then there is the claim that retractions are "skyrocketing." In a world of skyrocketing numbers of publications, this is not a meaningful claim. The relevant metric here is the retraction *rate*, so let's look at that. Steen et al. (2013) conclude that the retraction rate has increased since 1995, but that the overall retraction rate in the period 1973–2011 (based on an analysis of 21.2 million articles published in that interval) was 1 in 23,799 articles, or 0.004%.[32] Fang et al. (2012) conclude that the percentage of scientific articles retracted because of fraud has increased ~10-fold since 1975, but this still leaves the overall retraction rate at < 0.01%. It is difficult to see how this constitutes a general crisis in science.[33]

Moreover, it is not clear what the current increase in retraction rate means, because the notion and practice of retraction is a relatively recent one in the history of science. Historians have yet to study this matter closely, but it seems that the word "retraction" was until recently mostly used in the context of journalism.[34] According to Steen et al. (2013) the earliest retraction of a paper indexed in PubMed—the largest index of biomedical publications—was the 1977 retraction of a paper published in 1973.[35] To a historian, this relatively recent date is not surprising, insofar as faulty claims in science have traditionally been corrected by subsequent articles or ignored. Today we have claims of a retraction crisis, promulgated by websites such as "RetractionWatch.com," complete with social media outreach @retractionwatch and a Facebook page: https://www.facebook.com/retractionwatch/.[36]

RetractionWatch.com was founded in 2010, which suggests either that retractions have only of late become a problem or that they have only of late come to public attention. Here I am speculating, but I venture that the concept of retraction has gained traction in recent years because of heightened public scrutiny of science, which in turn has created conditions in which previously accepted practices of allowing erroneous claims to wither away are no longer considered adequate. If retractions were rare in the past but are common now, this may mean that science is more plagued by fraud or error. But it may simply signify that more people are watching, and mistakes that might in the past have been accepted as an unproblematic element of the progress of science are now being recast as unacceptable. In other words, for better or worse, it appears that we have changed our concept of what constitutes a problem in science.

The fact that most of Professor Krosnick's examples come from psychology and biomedicine is consistent with this interpretation. These are fields that generate a great deal of popular interest, and in which scientific results can have large social and commercial consequences. It is surely not a coincidence that the two studies he cites that found low rates of replication in biomedicine were undertaken by companies—Amgen and Bayer—with substantial financial stakes in scientific research outcomes. The competitive pressure of these high-stakes fields may indeed lead scientists to rush to publish work that turns out to be flawed. These are fields that are also heavily covered by mass media, who often run articles on single studies that may not be upheld by further work, leading perhaps to a biased impression of the overall state of science.

I can think of no example of a prominent retraction in geomorphology or paleontology.[37] There is, however, a recent highly publicized case in hydrology that merits consideration. A study published in a leading peer-reviewed journal found no effects on groundwater from hydraulic fracturing operations for gas production. The result garnered media attention because it seemed to undercut a major source of public concern about and source of opposition to fracking. However, a conflict of interest was later revealed: A gas company had partially funded the study, supplied the samples, and was involved in the study design, and one of the authors had worked for the company. The authors had not disclosed these potentially biasing factors.[38] The journal undertook a review of the situation, and invited me to write a paper on the necessity of financial disclosure (which I did).[39] Meanwhile, other researchers came to contrasting conclusions about the relations between proximity to gas wells and groundwater contamination.[40] We do not know which side in this debate is correct scientifically, but we do know that one side had a conflict of interest that could have affected their results.[41]

What do we conclude from all this? One obvious conclusion is that peer review is a highly imperfect process: bad and biased papers do get published. In the domain of endocrine-disrupting chemicals it has been shown that some published papers use strains of mice that are known to be insensitive to the effects under investigation.[42] Why would someone do that? It could be accidental—perhaps the researchers were unaware that these strains were insensitive—or it could be deliberate. It could also be that, knowing their funders' desires, researchers introduced a subconscious bias. Scientific papers are complex, and if the methods being used appear to be standard a reviewer might not examine them in detail. However, if reviewers are aware that the study's funders had a vested interest in a particular outcome, they may pay just a bit more attention.

We should also acknowledge that sometimes papers are *wrongly* retracted because of social or political pressure.[43] These cases may be rare or they may not be. The available evidence makes it difficult to judge. And this leads to another question: Is this is a global problem or not? The papers Krosnick cites were all published in English-language journals, which suggests that the lion's share were produced by English-speaking researchers or institutions. In the United Kingdom, recent changes in evaluation and funding of research universities have greatly increased pressures on scientists to increase their rate of publication. In the United States, funding rates have gone down dramatically compared to the 1960s, increasing the competitive pressure on researchers to produce results in a timely fashion in order to compete for the next round of funding. These factors contribute to pressure to produce—and not to take too much time checking results. One empirical test we might undertake would be to look at the country of origins of the researchers whose papers are being retracted.

There may well be general problems in contemporary science born of the competitive pressure to publish quickly and move onto the next fundable project, but Krosnick has not made the case. Despite his suggestion that problems are rampant throughout science, most of the evidence he offers involves a few domains and is drawn from English-language journals. This does not prove that all is well elsewhere, but Krosnick's argument slips from domains where problems are evident to domains where they are not.

This lack of clear and quantitative evidence permits him to make what I consider the least supported of his comments, that

"We don't need to know whether a project's funding comes from ExxonMobil or the National Science Foundation," and that "the source of funding . . . [is] among the least of our problems. A huge amount of research has been funded by federal agencies and private foundations that have no real agendas other than supporting scientists' making discoveries as quickly as possible." Here Krosnick makes both a logical and an empirical error. Logically, he succumbs to the fallacy of the excluded middle. Even if it were demonstrated that the replication problem was pervasive, it would not exclude the possibility of other serious problems in research. Empirically, we have strong empirical evidence of adverse effects when research is funded by self-interested parties.

It has been established that the tobacco industry long funded scientific research with the explicit goals of confusing the public, escaping legal liability by delaying epistemic closure, blocking public policy aimed at curtailing smoking, and, above all, maintaining corporate profitability by keeping smokers smoking.[44] By the judgment of nearly all scholars who have studied the matter, the industry succeeded. The link between tobacco and cancer was demonstrated by the 1950s, but smoking rates in United States began to decline dramatically only in the 1970s, when the tobacco strategy began to be exposed and thereby to become less efficacious.[45] While it is impossible to prove a counterfactual, the available evidence strongly suggests that if the tobacco industry had not interfered with scientific research and communication, more people would have quit smoking sooner and lives would have been saved.

The tobacco story is egregious, but not unique. Scholars have demonstrated the effects of motivated industry funding in the realms of pesticides and other synthetic chemicals, genetically modified crops, lead paint, and pharmaceuticals.[46] Recently, some have noted that a disproportionate amount of environmental research is now funded by the fossil fuel industry.[47] While the effects of the latter are not yet entirely evident, it seems reasonable to suppose that at minimum this is influencing the focus of research projects (such as emphasizing carbon sequestration as an answer to climate change versus energy efficiency, for example), and could be biasing the interpretation of scientific results.[48]

There is an additional problem that merits attention, one that increasingly makes it difficult for observers—or even scientists themselves—to differentiate legitimate from facsimile science. (By this term I mean materials that carry the accoutrements of science—including in some cases peer review—but fail to adhere to accepted scientific standards such as methodological naturalism, complete and open reporting of data, and the willingness to revise assumptions in the light of data.)[49] This is the problem of for-profit and predatory conferences and journals.

In recent years, various forms of sham science have proliferated. Some of them appear to be motivated purely for profit, charging substantial fees to attend their conferences or publish in their journals, fees that many scientists pay out of their research funds. Last year, one facsimile science institution run by a Turkish family was estimated to have earned over $4 million in revenue through conferences and journals.[50] Others may have disinformation as their intent, as they provide outlets for the tobacco, pharmaceutical, and other regulated industries to make poorly supported and false claims, and then insist that they are supported by "peer-reviewed science."[51]

A 2018 article "Inside the Fake Science Factory" discusses the findings of a team of researchers who analyzed over 175,000 articles published in predatory journals and found extensive evidence of published studies and conferences funded by major corporations, including the tobacco company Philip Morris, who have been found responsible in US courts for fraud based in part on their use of sham science to promote and defend their products.[52] Other participating companies, according to the report, included the pharmaceutical company AstraZeneca and the nuclear safety company Framatone. When the predatory journals publish these companies' research, the companies can claim that it is "peer reviewed," thus implying scientific legitimacy. But the damage spills over into academia, further blurring the boundary between legitimate and facsimile science: the researchers found hundreds of papers from academics at leading institutions, including Stanford, Yale, Columbia, and Harvard.[53] Whether the academic authors realize that they are publishing in sham journals is unclear; probably some do and some do not. The *New York Times* has called this phenomenon "fake academia." The phenomenon is sufficiently recognized that *Wikipedia* has an entry for "predatory conferences."[54]

Facsimile science can also be used by start-up companies to generate a supposedly scientific basis for proposed drugs and treatments, such as the company First Immune, which "had published dozens of 'scientific' papers in these predatory journals lauding the effectiveness of an unproven cancer treatment called GcMAF.... The CEO of First Immune, David Noakes, will stand trial in the United Kingdom later this year for conspiracy to manufacture a medical product without a license."[55]

No doubt these activities are bad for science, insofar as they can generate confusion within expert communities, but in many cases, experts will likely see the flaws in many if not most instances of facsimile science. The greater risk, I believe, is that to the extent that the public learns about these corrupt practices, they may come to distrust science generally. It is essential for academic scientists to pay attention to these issues, particularly the question of who is funding their science and to what ends, to insist in all circumstances of full disclosure of that funding, and to reject any grants or contracts that involve non-disclosure or non-publication agreements. In this sense, Professor Krosnick and I agree: It is essential for scientists to keep their house in order.

The difficulty of keeping our own house in order is underscored by one of the examples on which Professor Krosnick relies: the Amgen Pharmaceuticals replication study of papers published in *Science, Nature,* and *Cell.* These are top-flight journals that reject most of what is submitted to them and often boast about the importance of the work published in their pages, and, as Professor Krosnick notes, many scientists experience institutional pressure to publish in such prestigious journals. This may increase the odds that they exaggerate the novelty or significance of their results. *But how reliable is the Amgen study?*

In his note 18, Krosnick oddly cites not the Amgen study itself, but a very interesting and useful study of replication in psychology, which highlights the tension in science between innovation and replication and the need for both. "Innovation points out paths that are possible; replication points out paths that are likely; progress relies on both."[56] This paper discusses the Amgen study, albeit in passing. Here is what those authors have to say about the latter: "In cell biology, two industrial laboratories [Amgen and Bayer] reported success replicating the results of landmark studies in only 11 and 25% of the attempted cases .... These numbers are stunning but also difficult to interpret because no details are available about the studies, methodology, or results. With no transparency, the reasons for low reproducibility cannot be evaluated." Why didn't the Amgen scientists offer details about their study? We cannot say, because the published article was not, in fact, a peer-reviewed study, but a "Comment" by two authors, one an Amgen scientist and the other an academic.[57] It specifically address the problem of "suboptimal preclinical validation" in oncology trials, their recommendations equally specifically addressed to oncology research.[58]

Cancer is both a dreadful and a scientifically complex disease and the authors offer numerous reasons why promising early results may not translate into effective treatments. They also note that "it was acknowledged from the outset that some of the data might not hold up, because papers were deliberately selected that described something completely new." [59] In other words, the sample was deliberately and selectively focused on novel results; it was not a general appraisal of reproducibility in biomedicine. Admittedly, the replication rate achieved—11%— was very low. But was it "shocking?" Given that the papers were selected *because* they were novel and surprising, it strikes me as unsurprising that on further inspection most of them did not hold up. As I have stressed throughout this book, scientific knowledge consists of bodies of theory and observations. One paper does not constitute—*cannot* constitute—a scientific proof. If pharmaceutical companies design clinical trials based on inadequately verified scientific claims, that is certainly problematic, but it's not clear that the problem lies in *science*.

I agree with Professor Krosnick that suboptimal practices and problems in science need to be openly acknowledged and addressed; that is precisely the purpose of this book! But if we overgeneralize the problem, and are cavalier about funding (or any other kind of bias), it will be difficult if not impossible to assess either the extent or the cause of the replication crisis.

Professor Krosnick's comment underscores the need for the overall project of which this book is a small part: academic history and philosophy of science. He suggests that scientists rush to publish and exaggerate novelty because they "want to publish . . . counterintuitive findings" and, paradoxically, "don't want to admit that they didn't predict their findings." The first idea—that scientists should seek out counter-intuitive results that will upset the applecart of received wisdom—was a central idea for Karl Popper. The second idea—that we should be able to predict our findings—is the centerpiece of the hypothetico-deductive model of science. In chapter 1 we saw that both these models have serious logical flaws and neither works well as an accurate empirical description of scientific activity. If Professor Krosnick is right about what scientists want, then scientists are wanting a lot of wrong things. In that case, I hope that this book will help them to appreciate what science can and cannot give them.

AFTERWORD

Truthiness. Fake news. Alternative Facts. Since these Princeton Tanner Lectures were delivered in late 2016, the urgency of sorting truth from falsehood—information from disinformation—has exploded into public consciousness.[1] Climate change is a case in point. In the United States in the past two years, devastating hurricanes, floods, and wildfires have demonstrated to ordinary people that the planetary climate is changing and the costs are mounting. Denial is no longer just pig-headed, it is cruel. The American people now understand—as people around the globe have already for some time—that anthropogenic climate change is real and threatening.[2] But how do we convince those who are still in denial, among them the president of the United States, who has withdrawn the United States from the international climate agreement and declared climate change to be a "hoax"?[3]

Moreover, on many other issues our publics are as confused as ever. Millions of Americans still refuse to vaccinate their children.[4] Glyphosate pesticides remain legal and widely used, even as the evidence mounts of their harm.[5] And what about sunscreen?

In this social climate, one might conclude that the arguments of this book are overly academic, that the social and political challenges to factual knowledge are so great that we should be focused on these dimensions and not on epistemology. As the co-author of *Merchants of Doubt*—a book dedicated to explicating ideologically motivated opposition to scientific information—I might be expected to do just that. That would be a mistake.

As Erik Conway and I showed in that book, the core strategy of the "merchants of doubt" is to create the impression that the relevant science is unsettled, the pertinent scientific issues still appropriately subject to contestation. If we respond on their terms—offering more facts, insisting that these facts *are* facts—then they win, because now there *is* contestation. When it comes to doubt-mongering, one cannot fight fire with fire. One has to shift the terms of debate. One way to do so is by exposing the ideological and economic motivations underlying science denial, to demonstrate that the objections are not scientific, but political. Another is by explaining how science works and affirming that, under many if not all circumstances, we have good reason to trust settled scientific claims. In *Merchants of Doubt,* Conway and I did the first. Here, I am attempting to do the second.

The argument of this book is that the answer to our question— Why Trust Science?—is not that scientists follow a magic formula ("the scientific method") that guarantees results. That idea persists in textbooks and the popular imagination, but it does not stand up to historical scrutiny. What does stand up is a portrait of science as a communal activity of experts, who use diverse methods to gather empirical evidence, and critically vet claims deriving from it.

The diverse methods of science have identifiable common elements. One is experience and observation of the natural world; another is the collective critical scrutiny of claims based on those experiences and observations. In chapter 1 we developed the argument that the appropriate basis for lay trust in science is the sustained engagement of scientists with the natural world, coupled with the social character of science that includes procedures for critical interrogation of claims.

All social arrangements rely on trust, and many involve expertise, be it from doctors, dentists, plumbers, electricians, car mechanics, accountants, auditors, tax attorneys, real estate appraisers, or what-have-you. Even buying a pair of shoes may rely on trusting the salesman to measure our feet properly. If trust in experts were to come to a halt, society would come to a halt, too. Scientists are our experts in studying the natural world and sorting out complex issues that arise in it. Like all experts, they make mistakes, but they have knowledge and skills that make them useful to the rest of us. The crucial component that separates science (and here I include the social as well as the natural sciences) from, for example, plumbing, is the centrality of the social vetting of claims.

The critical scrutiny of scientific claims is not done individually; it is done collectively, in communities of highly trained, credentialed experts, and through dedicated institutions such as peer-reviewed professional journals, specialist workshops, the annual meetings of scientific societies, and scientific assessments for policy purposes.[6] A crucial aspect of this process is *revision*: most peer-reviewed papers are revised many times prior to publication, both informally as preliminary results are presented at conferences and workshops and drafts are sent to colleagues for comment, and then formally through editorial peer review. Papers are then revised in response to reviewers' suggestions of clarifications and corrections. If errors are detected after publication, journals may issue errata or retractions. (In this regard, retractions should be seen as essentially a good thing.) This process of critical scrutiny and revision is what philosopher Helen Longino has called "transformative interrogation," what anthropologist Bruno Latour calls the "agonistic field." It is the process by which, as historian Martin Rudwick has stressed, novel solutions to problems are developed, accepted, and sustained as *facts*.[7]

Exchanges among scientists at times get testy, but this is to be expected when hard-won intellectual accomplishments are called into question. The fact of contestation—even highly emotional contestation—is not by itself evidence that anything is wrong. (On the contrary, it may be a sign that things are right, as scientists are taking a challenge seriously and neither ignoring nor dismissing it.) Through this process of contestation, novel claims come to be intersubjectively accepted and ultimately viewed as objectively true. The social aspect of scientific work is thus crucial to the question of whether or not scientific conclusions are warranted, because it helps to ensure that conclusions are not merely the opinions of individuals or dominant groups, but something less personal and more reliable. A claim that has survived critical scrutiny becomes established *fact*, and collectively the body of established facts constitute scientific *knowledge*.

The beauty of this picture is that we can now explain what might otherwise appear paradoxical: that scientific investigations produce both novelty and stability. New observations, ideas, interpretations, and attempts to reconcile competing claims introduce novelty; critical scrutiny leads to collective decisions about what obtains in the world and hence to stability of knowledge claims. This picture also helps us to appreciate the irony that what was once viewed as an attack on science—the articulation of its social character—provides the basis of the strongest defense we can make of it.[8]

That said, those of us who wish to defend science from ideologically and economically interested attack must be not only willing and able to explain the basis of our trust in science, but also to understand and articulate its limits. This means coming clean about the various ways in which things can go wrong. In chapter 2, we explored a number of instances where scientists, in hindsight, did get things wrong. Here, we saw the salience of three matters especially: 1) consensus, 2) diversity, and 3) methodological openness and flexibility.

Consensus is essential to our argument for the simple reason that we have no way to know *for sure* if any particular scientific claim is true. As philosophers going back to Plato (and perhaps before) have long recognized, we do not have independent, unmediated access to reality and therefore have no independent, unmediated means to judge the truth content of scientific claims. We can never be entirely *positive*. Expert consensus serves as a proxy. We cannot know if scientists have settled on the truth, but we can know if they have settled. In some cases where it is alleged in hindsight that scientists "got it wrong," we find on closer examination that there was, in fact, no consensus among scientists on the matter at hand. Eugenics is a case in point.

Diversity is crucial because, *ceteris paribus*, it increases the odds that any particular claim has been examined from many angles and potential shortcomings revealed. Homogenous groups often fail to recognize their shared biases. In chapter 2, we saw not only how the Limited Energy Theory instantiated prevailing late nineteenth-century American gender bias, but also how Dr. Mary Putnam Jacobi shone a light on those biases and in doing so revealed serious flaws in both the theory and its evidentiary basis. We also saw how socialist geneticists were particularly articulate in their opposition to eugenics, drawing on their politics to question the obvious class bias in many eugenic theories and proposals. One did not have to be a socialist to question eugenics, but socialist class consciousness played a role in a substantial line of dissent.

Methodological openness and flexibility are necessary because when scientists become rigid about method, they may miss, discount, or reject theories and data that do not meet their standards. We saw this at play in the history of continental drift theory, as American scientists rejected a theory that did not follow their preferred inductive approach; in the history of the contraceptive pill, where gynecologists rejected case reports from patients because they were viewed as subjective and therefore unreliable; and in the evaluation of dental floss, where double-blind trials were simply not possible.

These insights make clear that we are not powerless to judge contemporary scientific claims. We can ask: Is there a consensus? Is the community undertaking the studies diverse, both demographically and intellectually? Have they considered the issue from a variety of perspectives? Have they been open to diverse methodological approaches? And have they paid attention to all the relevant evidence, not missing or discounting some substantial portion of it? Have they avoided becoming fetishistic about method?

In closing, let's consider one more topic: sunscreen. It is well known that some widely used ingredients in sunscreens—particularly oxybenzone—may disrupt endocrine function in laboratory animals.[9] Oxybenzone is also is toxic to corals.[10] The state of Hawaii has banned the sale of oxybenzone-bearing sunscreens, and many consumers (myself included) have switched to mineral-based formulations.[11] However, recently some scientists and physicians have questioned the use of sunscreen, *tout court*, and in January 2019, *Outside* magazine reported on new evidence suggesting that conventional wisdom on the benefits of using sunscreen was wrong.

The article focused on "rebel" dermatologist Richard B. Weller, who believes that sunlight lowers blood pressure, which in turn lowers the risk for heart disease and stroke—two of the biggest killers in the industrialized world. If Weller is right, then the widespread, habitual use of sunscreen may have adverse health effects. In the article's headline, the magazine demanded provocatively: "Is Sunscreen the New Margarine?"[12]

The argument begins with the established correlation between sunshine and heart health. As the article reported, "high blood pressure, heart disease, stroke, and overall mortality all rise the farther you get from the sunny equator, and they all rise in the darker months." But is sunshine the controlling factor? After all, food is often better in Mediterranean climes than in high latitudes (think Italy vs. Norway), and people eat more fresh fruits and vegetables and typically get more exercise in summer. Or perhaps life is more stressful when you have to deal with snow and ice and long, dark winter nights. However, at least one controlled study suggests that the causal factor *is* sunshine: when volunteers were exposed to the equivalent of thirty minutes of summer sunlight (without sunscreen), their blood pressure decreased. Moreover, there is a known mechanism to explain this relationship: nitric acid in the blood dilates blood vessels and thereby lowers blood pressure, and sun exposure increases blood nitric acid. Thus: sun exposure increases nitric oxide, which decreases blood pressure, which reduces the risk of heart attack and stroke. Not bad for something that is readily available to most of us, free of charge. So ditch the sunscreen and head outside, right? That's what the *Outside* magazine writer concluded, wondering "How did we get it so wrong?"

Did "we" get it wrong? More to the point, did scientists (or physicians) get it wrong? If you only read this article, you would conclude that they did. The American Academy of Dermatologists, for example, advises "everyone" to use sunscreen, to routinely seek shade during the hours of 10 a.m. and 2 p.m., to wear protective clothing including long-sleeved shirts, pants, hats, and sunglasses, and to get Vitamin D through diet. "Don't seek the sun," they state without qualification.[13] The *Outside* article calls this a "zero-tolerance" stand.

However, there are a number of problems with the magazine's conclusion. The article relies heavily on a study by Dr. Weller that has not yet been published. "Weller's largest study is due to be published later in 2019," we are told. Perhaps the study will be game-changing, but until it goes through peer review and is published, we are in no position to judge—and neither is *Outside* magazine.

Weller is a coauthor on two papers that have been published—one in 2014 and one in 2018. Both relied on very small samples: twenty-four (eighteen men and six women) and ten participants (all male), respectively. No matter what they found, it would be unwise to summarily reject a huge body of existing science demonstrating the adverse effect of sun exposure (skin cancer) on the basis of such small studies.

Moreover, what the studies found does *not* support the conclusion that the magazine article asserts.

The 2014 paper found a small, transient decrease in diastolic blood pressure (e.g., from 120 to 117) associated with an exposure to artificial UVA light equivalent to thirty minutes in a Mediterranean region. The authors assert the significance of this result, arguing that "any amount of BP [blood pressure] reduction is protective against stroke and cardiovascular mortality . . . and the magnitude of changes observed in this study would appear to be large enough to account for the standardized mortality differences in populations living at different latitudes." That might be true if the observed changes were sustained, but temporary blood pressure effects from nitrates are not strongly associated with long-term improved cardiovascular health.[14] Unless people are staying outside *a lot*, the significance of this finding is unclear. Certainly, it is far from demonstrated.

The 2018 paper found a transient effect on blood nitric acid levels and resting metabolic rate, but *no effect at all on blood pressure*. This immediately calls into question the alleged mechanism. *Outside* implied that the mechanism was known, but in fact it was a hypothesis that these studies were designed to test, and this test did not confirm it! Moreover, if something is a causal factor, we expect to find a dose-response relationship: more of the cause should produce more of the effect. The study found no dose-response relationship, forcing the authors to admit that their findings "contrasted" with their hypothesis. And both studies involved artificial, UVA-only light, leaving it unclear how well this any of this correlates with natural sun exposure.

Dr. Weller may one day be proven right, but at present the alleged benefits of sunlight on blood pressure are not even close to proven. In contrast, the connection between sun exposure and skin cancer is.[15] This is why dermatologists are advocates of sunscreen and sun avoidance, particularly for fair skinned people in Europe, North America, Australia, and New Zealand. Sun exposure can lead in the short run to painful sunburn, and in the long run to early aging of the skin and skin cancers, including deadly melanomas. The scientific evidence for this is abundant and well established.

If we look at the guidelines offered by leading organizations of dermatologists in the United States, United Kingdom, and Australia, we do find some subtle differences of opinion and emphasis. In contrast to the American "zero-tolerance" stand, the Australian Cancer Council discusses the risks and benefits of sun exposure, providing "guidance on how much sun you need and how to protect yourself from getting too much."[16] They advise protection (hats, sunglasses, and sunscreen) when UV levels exceed three, which in most cases equates to protection in summer, but not winter.[17] (This contrasts with the dominant advice in the United States, which is to use sunscreen year-round.) The argument in favor of some sun exposure is not, however, the effect on blood pressure, but on Vitamin D.

"A balance is required between avoiding an increase in the risk of skin cancer by excessive sun exposure and achieving enough exposure to maintain adequate vitamin D levels."[18]

The British Association of Dermatologists also advocates a balanced approach:

> Nobody wants to spend the entire summer indoors, and indeed some sunshine, below sunburn level, can be good for us, helping the body to create vitamin D and giving many of us a feeling of general wellbeing as we enjoy outdoor summer activities.
>
> However, all too often we over-do our sun exposure which can lead to a range of skin problems, the most serious of which include skin cancer. Other summertime skin problems include sunburn, photosensitive rashes and prickly heat. In addition, sun exposure can worsen already existing conditions like rosacea.[19]

The UK dermatologists stress the differences among people, noting that light-skinned people burn more easily and therefore need more protection than darker people. However, in the end, their advice (at least for the light-skinned among us) is more or less the same as the Americans': protect yourself with a hat, clothing, and sunglasses, use sunscreen of at least SPF 30 on exposed skin, and stay in the shade at midday. Lest they be accused of being old-fashioned, they also suggest getting the World UV app, which provides "real time information on daily UV levels across over 10,000 locations across the globe."[20]

Where does this leave us? While there are some differences of opinion among dermatologists on how to balance the risks of skin cancer (and other forms of skin damage) with the benefits of sun exposure (Vitamin D metabolism), overall, physicians have a consensus on the benefits of protecting yourself from the sun. Scientists did not get this wrong, *Outside* magazine did.

Of course, there may well be benefits to sun exposure that go beyond Vitamin D. Californians don't need British doctors to tell them that being out in the sun makes them feel good, and there is clearly a reason people vacation in sunny places. Moreover, dermatologists—with their focus on protecting the skin from the sun's ill effects—might be slow to attend to evidence that some degree of sun exposure is good for you. And it is interesting that American dermatologists seem to take a harder line than British and Australian doctors. But then Americans take a harder line than Australians on many things.

Good decision-making requires integration of information. Being healthy involves more than just avoiding carcinogens.[21] It also involves relaxation, recreation, stress reduction, and many other things that Europeans and Australians seems to be better at than Americans, and to which science has been a bit slow to attend. There is not only more on heaven and Earth than is dreamed of in our philosophies, but also more than is understood by our sciences.

There is much we do not know, but that is no reason not to trust science on the things we do know. The argument for trust in science is not an argument for blind or blanket trust. It is an argument for warranted confidence against unwarranted skepticism in scientists' findings in their domains of expertise.

NOTES

Introduction

1. As Naomi Oreskes has argued in a previous book coauthored with Erik M. Conway, *Merchants of Doubt,* and a documentary film of the same name.

2. See *CBS Evening News,* January 30, 2019, https://www.cbsnews.com/video /how-long-will-the-cold-snap-last/; and *PBS Newshour,* January 30, 2019, segment with Dr. Jennifer Francis, https://www.pbs.org/newshour/show/why-the-midwests -deep-freeze-may-be-a-consequence-of-climate-change.

3. Princeton University Press secured anonymous expert reviews of an earlier version of the manuscript, and their detailed and helpful comments have been of great help in revising and improving the manuscript.

4. As I once heard the political philosopher Joseph Cropsey say, at a seminar organized by Harvey C. Mansfield in the Harvard Government Department: while the greatest minds rise above many of the prejudices of their times, no one rises above all the prejudices of their times.

Chapter 1. Why Trust Science? Perspectives from the History and Philosophy of Science

1. I hesitate to use the word crisis, but on the other hand, rejection of the science of vaccines is a matter of life and death, and rejection of climate science has now become so.

2. This claim was picked up by various media outlets, including several promoting conspiracist ideation. Jones, "About Alex Jones."

3. Mnookin, *The Panic Virus.*

4. Miller, *Only a Theory*; "Evolution Resources from the National Academies."

5. Newport, "In U.S., 46% Hold Creationist View of Human Origins."

6. National Center for Science Education, "Background on Tennessee's 21st Century Monkey Law."

7. On the history of attempts to teach creation in the classroom, see Minkel, "Evolving Creationism in the Classroom." On the broader history of American creationism, see Larson *Summer for the Gods*; Numbers, *The Creationists*; Michael Berkman and Eric Plutzer, *Evolution, Creationism and the Battle to Control America's Classrooms*.

8. See Zycher, "The Enforcement of Climate Orthodoxy and the Response to the Asness-Brown Paper on the Temperature Record"; Hayward, "Climategate (Part II)"; Sample, "Scientists Offered Cash to Dispute Climate Study"; Union of Concerned Scientists, "Global Warming Skeptic Organizations"; and Sachs, "How the AEI Distorts the Climate Debate."

9. Sachs, "How the AEI Distorts the Climate Debate."

10. Zycher, "Shut Up, She Explained."

11. Richards, "When to Doubt a Scientific 'Consensus.'"

12. This is not to suggest that the authority of science has never been questioned. Certainly the values of science have been questioned by many writers, poets, religious leaders, and others. Mary Shelley's indictment of scientific hubris in her classic work, *Frankenstein*, particularly comes to mind, along with Goethe's *Faust* and other variations of the Faust legend. Various artists and poets who implicitly or explicitly criticized science on diverse grounds, including the disenchantment of nature (see, for example, Harrington, *Reenchanted Science*, 1999). My point here is that as a source of authority on *empirical* questions, science has been broadly accepted in recent Western culture, which is, in part, why the current state of affairs seems to many of us to be so shocking.

13. A particularly cogent refutation of this strategy is found in Bloor, *The Enigma of the Aerofoil*. See also my discussion in *Rejection of Continental Drift*, pp. 313–18.

14. Shapin, *A Social History of Truth*. See also discussion in Frodeman and Briggle, "When Philosophy Lost Its Way."

15. Crosland, *Science under Control*. It is also a reason, although not the only one, why women were generally excluded.

16. Bourdeau, "Auguste Comte."

17. For background on the rise of secularism in the nineteenth century, see Weir, *Secularism and Religion in the 19th Century*.

18. Comte, *Introduction to Positive Philosophy*, on p. x.

19. Ibid., p. 2.

20. Morris and Brown, "David Hume."

21. Comte, *Introduction to Positive Philosophy*, p. 4.

22. Ibid., pp. 4–5.

23. Ibid., p. 23.

24. Emphasis added. Thus we find that Bruno Latour is in fact a positivist. Ibid.

25. Note that Comte does not take this to logical conclusion regarding gender. Bourdeau, "Auguste Comte."

26. Richardson and Uebel, *Cambridge Companion to Logical Empiricism*, use the terms logical positivist and logical empiricist (and sometimes neopositivist) interchangeably, noting that while some philosophers in the mid-century thought these terms had different referents most did not, and that by the 1930s logical empiricist was the term preferred by most discussants.

27. Ayer, *Language, Truth and Logic*, p. 13.

28. Ibid., p. 11.

29. Friedman and Creath, *The Cambridge Companion to Carnap*; Quine and Carnap, *Dear Carnap, Dear Van.*

30. I focus here on challenges relevant to the philosophy of science. There were also substantive challenges in the domain of mathematics, e.g., the attempts by Bertrand Russell and A. N. Whitehead to place mathematics on a logical footing, but this is beyond my expertise and ambition.

31. Popper's critical rationalism is directly related to his politics: indeed, he insists, throughout his work, that his project is both epistemological and political: he believes that the sort of skeptical attitude necessary for scientific work is the same as what is necessary to resist authoritarianism.

Both his politics and his epistemology are radically individualistic; *Conjectures and Refutations* is dedicated to von Hayek. Perhaps for this reason his work was widely taken up by anti-Communists in Eastern Europe, as well as by neo-liberals. See Mirowski and Plewe, *Road from Mt. Pelerin.*

32. Popper, *Conjectures and Refutations*, p. 46ff.

33. Popper sometimes expanded his positions in ways that softened them. So as noted above, his theory seems to be radically individualistic, insofar as he focuses on the attitude of the individual scientist. But, on the other hand, he also notes that the objectivity of the scientist does not, in fact, reside in the individual, but in the objective nature of scientific theory insofar as a theory must be communicated to others for it to be subjected to severe tests. For example, in *The Myth of the Framework*, he explicitly rejects the idea that rational discussion in a community is impossible unless they "share a common framework of basic assumptions," or at least have agreed upon such a framework. But then he allows that there is a kernel of truth in this myth, namely, that fruitful and rational discussion "among participants who do not share a common framework may be difficult." Either way, here he is acknowledging that scientific discussions take place among communities (Popper, *Myth of the Framework*, 34–35). Put another way: theories are not tested only by the individuals who perform the test, but also by the community of experts to whom those tests are reported. Helen Longino makes a similar point in *Fate of Knowledge*, 5–7, when she

notes that the process of refutation, which is key to Popper's concept of science as "conjecture and refutation," acknowledges the role of other scientists' criticisms in causing us to rethink our views. Thus even for Popper, criticism—which is central to science—is a social activity. Put another way, if we take criticism seriously, then we see that the social component of science is not epiphenomenal, but constitutive.

34. Sady, "Ludwik Fleck." See also Löwy, *The Polish School of Philosophy of Medicine.*

35. Fleck, "Scientific Observation and Perception in General."

36. Fleck and Kuhn, *Genesis and Development of a Scientific Fact,* p. 42.

37. Ibid.

38. Longino, *Fate of Knowledge,* p. 122.

39. Fleck identified the problem of the isolation of the expert from the non-expert community. The expert is "already a specially modelled individual who can no longer escape the bonds of tradition and of the collective; otherwise he would not be an expert" (Sady, sec. 7). Public presentation of science takes the fluid and interactionist reality of science and presents it as a fixed and finished project, making science seem more certain and dogmatic than it actually is.

40. Together with the American chemist J. Willard Gibbs, Duhem developed the mathematics that describes the relationship between changes in the chemical potential of substances in a system and changes in the temperature and pressure of the system—something I stayed up many long nights studying in my days as a geochemist.

41. The French original may be accessed at https://archive.org/stream /lathoriephysiquoounkngoog#page/n6/mode/2up.

42. De Broglie, forward, in Duhem, *The Aim and Structure of Physical Theory,* p. xi.

43. Ibid., p. 220.

44. Ibid., p. 219.

45. Here, he is attempting to distinguish between experimental laws as regularities, like F=Ma, and the explanatory theory that makes sense of them, such as the laws of motion.

46. Duhem, *The Aim and Structure of Physical Theory,* p. 180.

47. *Aim and Structure* was published in 1906, but according to de Broglie, Duhem wrote it in 1905, when Einstein published his work on the photoelectric effect. This result may have been in Duhem's mind.

48. Duhem, *The Aim and Structure of Physical Theory,* p. 183.

49. Ibid., p. 185.

50. Ibid., p. 187.

51. Ibid., p. 180.

52. Ibid., p. 181.

53. This was an accusation that in the 1920s would be made against Alfred Wegener, see Oreskes, *Rejection of Drift*.

54. Duhem, *The Aim and Structure of Physical Theory*, p. 217.

55. Ibid., p. 212.

56. Ibid. p. 270. So in the end he does seem to privilege theory over experiment, but this is beyond the scope of this chapter. The key point is that it is history that gives us grounds for confidence, in the long run.

57. Quoted in Zammito, *A Nice Derangement of Epistemes*, p.17. Note this makes clear he's not doubting the existence of the external world; the issue is how we respond to evidence from it.

58. Quine, "Two Dogmas of Empiricism." Quine also emphasized what has come to be known as the "theory-ladenness" of observation. Duhem stressed that there are no experiments without instruments, and there are no instruments without theory: "without theory it is impossible to regulate a single instrument or to interpret a single reading." Quine pushes this further to argue that without theory there is no observation. All observations are created and interpreted in the framework of pre-existing theory, and thus observation has no life of its own.

59. Zammito, *A Nice Derangement of Epistemes*, p. 20.

60. Ibid.

61. Conant, *Harvard Case Histories in Experimental Science Volume I*.

62. Fuller, *Thomas Kuhn: A Philosophical History for Our Times*; Reisch, "Anticommunism, the Unity of Science Movement and Kuhn's Structure of Scientific Revolutions"; Galison, "History, Philosophy, and the Central Metaphor."

63. Fleck and Kuhn, *Genesis and Development of a Scientific Fact*. I think this is a very important point—and it is not simply that the loner is more likely to be viewed as a crank than a maverick, he is more likely to *be* a crank.

64. Kuhn, *Reflections on My Critics*, on p. 247.

65. As an undergraduate I read *The Structure of Scientific Revolutions* with a group of friends—aspiring scientists all—and we liked it because it seemed realistic. Kuhn's description of scientists not questioning the larger assumptions under which they operated seemed true of our professors.

66. Kuhn himself denied this, and spent much of his later life in philosophy of language, attempting to sort the problems of scientific translation as a species of the general problem of translation.

67. Lakatos, *Criticism and the Methodology of Scientific Research Programmes*, on p. 181.

68. Kuhn and Conant, *The Copernican Revolution*, p. 182.

69. One student of mine asked how Kuhn's views are different from Fleck. As a historical matter, Kuhn had far more influence in Anglophone circles than Fleck;

from the US perspective Fleck has been rediscovered in recent years (e.g., Harwood, *Ludwik Fleck and the Sociology of Knowledge*, 1996). From the European perspective, one might argue that Kuhn borrowed heavily from Fleck without adequate acknowledgement. But Kuhn borrowed heavily in general; *Structure* does not have a very extensive bibliography. Mosner, *Thought Styles and Paradigms*, has recently argued that scholars have been too quick to equate their philosophies. It seems to me the obvious major difference involves Fleck's view of the evolution of ideas, which lack the abrupt disjuncture that Kuhn insists characterizes scientific revolutions.

70. Zammito, *A Nice Derangement of Epistemes*. One interesting question, which needs to be explored more, is the extent to which sociologists were inspired by Peter L. Berger and Thomas Luckman's *The Social Construction of Reality*. This book, published in 1966, is hard to interpret because the authors deliberately omit the names of prior scholars (see *Social Construction of Reality*, p. vi) as disruptive to the argument. However, they do acknowledge the influence of Alfred Schutz, an Austrian philosopher with links to one of the founders of neo-liberalism, Ludwig von Mises. Zammito (pp. 124–25) places Berger and Luckman in the tradition of American pragramtist George Herbert Mead, and suggests it had rather little influence on sociology of knowledge. He argues that social constructivism in science studies was much more a response to the Frankfurt school.

71. Barnes, *Interests and the Growth of Knowledge*.

72. Bloor, *Knowledge and Social Imagery*, p. 7.

73. Shapin and Schaefer, *Leviathan and the Air-Pump*, p. 332.

74. Sokal, *Beyond the Hoax*; Gross and Levitt, *Higher Superstition*; Gross, Levitt, and Lewis, *The Flight from Science and Reason*.

75. Barry Barnes, quoted in Zammito, *A Nice Derangement of Epistemes*, p. 134.

76. See Zammito, *A Nice Derangement of Epistemes*, and Hacking, *The Social Construction of What?* The term "social construction" is generally credited to Berger and Luckmann, *The Social Construction of Reality*.

77. Barnes, *Scientific Knowledge and Sociological Theory*, p. vii.

78. Zammitto, *A Nice Derangement of Epistemes*, p. 52.

79. Bloor, *The Enigma of the Aerofoil*, conclusion.

80. On this point, see my critique of Miriam Solomon: Oreskes, "The Devil Is in the (Historical) Details."

81. Feyerabend, *Against Method*, pp. 18–19. See also Motterlini (ed.), *For and Against Method*.

82. It is worth noting that the argument about the benefits of diversity in producing creative and effective outcomes is now widely accepted in the business community. See for example, Page, *The Diversity Bonus*, and Lowery, "Why Gender Diversity on Corporate Boards Is Good for Business."

83. David Bloor nicely revisits this point in his wonderful and underappreciated book, *The Enigma of the Aerofoil.*

84. Feyerabend, *Against Method*, p. 5.

85. Latour, *Science in Action.*

86. See also Galison and Stump, *Disunity of Science.*

87. I made a claim more than twenty years ago that the dream of positive knowledge had ended; John Sterman pointed out that it lived on in economics (Oreskes et al., 1994, *Verification, Validation, and Confirmation of Numerical Models in the Earth Sciences*, and Sterman 1994, *Letter*.) See also Ladyman et al., *Every Thing Must Go.*

88. Weinberg, *Facing Up.* In fact, it illustrates an important point to which we will return: Scientists should not be trusted when they venture outside their domain of expertise. Weinberg is a brilliant man. He won the Nobel Prize in 1979 for one of the most important developments of twentieth-century physics. But this comment reflects either a shocking ignorance of the history of science, or a shocking disregard of evidence compiled from another field. Either way it demonstrates that expertise is not transferable. We should trust Weinberg about physics, but not about its history.

89. It is important to note that nearly all the major feminist philosophers of science of this period (e.g., Evelyn Fox Keller, Ruth Hubbard, Scott Gilbert, Anne Fausto-Sterling, and maybe even Donna Haraway?) rejected the idea that their critique of science implicated them in ontological relativism. Certainly Keller, Hubbard, and Fausto-Sterling, who were themselves scientists, were interested (like Longino and Harding) in making a better, less biased, more objective science. See, for example, Keller, *Reflections on Gender and Science*, Hubbard, *Politics of Women's Biology*, and Fausto-Sterling, *Myths of Gender.*

90. An implicit assumption here is the demographic diversity will carry with it intellectual diversity. I address this in chapter 2.

91. My student Charlie Tyson raises an interesting point in regard to the issue of objectivity, and the criticism that leftish scholars including Harding endured because of their "relativist" positions on objectivity. The conservative intellectuals and media activists of the mid-twentieth century who sought to create their own journals to promote conservative views—men like William F. Buckley, for example, who found themselves shut out of some conversations because of the extremity of their views—did not simply charge that the mainstream media were biased. They rejected the concept of objectivity itself, or at least the conflation of objectivity with impartiality, and accepted their bias as a legitimate one. The mission statement of the publication *Human Events*, one of the pillars of early conservative media activism, is telling. "*Human Events* is objective; it aims for accurate representation of the facts. But it is not impartial. It looks at events though the eyes that are biased in favor of limited constitutional government, local self-government, private enterprise, and individual

freedom" (Hemmer, p. 32). These media activists thus introduced bias and partiality as legitimate values in reporting. Hence the irony of conservatives today who accuse the media and universities of "liberal bias" (Hemmer, p. xii). So it is simply incorrect to equate the questioning of standard views of objectivity with "the academic left," as Gross and Levitt did in their book, *Higher Superstitions*. Indeed I would argue that right-wing critics like Buckley rejected objectivity, while left-wing critics like Harding sought to improve it.

92. Harding, "Women at the Center." For one characteristic exchange, see Hicks, "Is Newton's *Principia* a Rape Manual?". On conservative responses to academic feminism see Schrieber, *Righting Feminism: Conservative Women and American Politics*. In hindsight Harding allows that *The Science Question in Feminism* had an "us v. them" tone, which she would not take today. See Flores, "Beyond the Secularism Tic—An Interview with Feminist Philosopher Sandra Harding." However, if the point of being provocative is to provoke, she certainly did that.

93. Longino, *Science as Social Knowledge*, 79; Harding, *The Science Question in Feminism*; Solomon, *Social Empiricism*.

94. Bernard, *An Introduction to the Study of Experimental Medicine*.

95. It is important to note that simply adding one woman or person of color to an otherwise homogeneous community will not solve the problem, as that isolated individual will likely not feel sufficiently secure to challenge the dominant worldview.

96. Longino, *Science as Social Knowledge*, p. 216.

97. Ibid., p. 80. See also my discussion of the Limited Energy Theory in chapter 2.

98. Longino, *Science as Social Knowledge*.

99. Ibid.

100. This is a theoretical rather than empirical argument; Longino did not have empirical evidence to support it in part because at the time she was writing women were only just regaining the positions in academic science that they had held and lost earlier in the century (see Rossiter, *Women Scientists in America*). Londa Schiebinger has given examples of how women in science have helped to open up new areas of investigation and offer alternative (better) theories in a number of domains (Schiebinger, *Has Feminism Changed Science*), but this work and any attempt to demonstrate that diverse scientific communities produce better theories suffers from the dilemma that in science we have no accepted metric of "better." Studies of diversity in business clearly show that diverse teams perform better by many standards, so much so that the finding now has a moniker: "The Diversity Bonus." It is now routine for people in the business community to argue that diversity is not just morally right, but profitable (see Page, *The Diversity Bonus*).

101. See note 89.

102. Ibid., p. 79.

103. Smithson, "Social Theories of Ignorance," in Proctor and Schiebinger, *Agnotology*. See also Giddens, *Consequences of Modernity*.

104. Oreskes et al., "Viewpoint," p. 20.

105. Oreskes, "Why We Should Trust Scientists."

106. Jon Krosnick, this volume, questions this assumption; I return to it in the discussion section.

107. Longino, *Fate of Knowledge*, pp. 106–7.

108. Yearley et al., "Perspectives on Global Warming."

109. On the idea of partisanal and non-partisanal knowledge, see Staley, "Partisanal Knowledge: On Hayek and Heretics in Climate Science and Discourse."

110. Oppenheimer, Jamieson, Oreskes, et al., *Discerning Experts*.

111. Laland et al., "The Extended Evolutionary Synthesis"; Laland et al., "Does Evolutionary Theory Need a Rethink?"

112. Laland, "What Use Is an Extended Evolutionary Synthesis?"

113. Part of the issue involves the question of whether evolutionary theory is in crisis, and therefore EES is a new paradigm, or not. Kevin Laland says that it is not; philosopher John Dupre says it is. See for example: Coyne, "Another Philosopher Proclaims a Nonexistent 'Crisis' in Evolutionary Biology."

114. Oreskes, *The Rejection of Continental Drift*.

115. Neumann, "Can We Survive Technology?" Saxon, "William B. Shockley, 79, Creator of Transistor and Theory on Race."

116. Redd, "Werner von Braun: Rocket Pioneer."

117. Laura Stark notes that it is often assumed that professional experience and knowledge translates into "rare abilities to judge the quality, veracity, or ethics of knowledge outside of research settings." She does not explicitly state that this assumption is incorrect, but ample evidence from the history of science and her own work on Institutional Review Boards supports the conclusion that it is. See Stark, *Behind Closed Doors*, p. 31.

118. On lay expertise, see, for example, Epstein, *Impure Science*.

119. Mohan, *Science and Technology in Colonial India*.

120. Goonailake, "Mining Civilizational Knowledge."

121. Ellis et al., "Inpatient General Medicine Is Evidence Based"; Ernst, "The Efficacy of Herbal Medicine—an Overview."

122. Goonatilake, "Mining Civilizational Knowledge."

123. Scott, "Science for the West."

124. Semali and Kincheloe, *What Is Indigenous Knowledge?*; Schiebinger and Swan, *Colonial Botany*. An excellent discussion of the general issue of how to understand indigenous knowledge as science, both in general and in the specific case of Cree hunting practices, is found in Scott, "Science for the West." For a

problematization of the issue, see Agrawal, "Dismantling the Divide between Indigenous and Scientific Knowledge."

125. Walker, "Navigating Oceans and Cultures."

126. Conis, "Jenny McCarthy's New War on Science"; Campbell, "The Great Global Warming Hustle."

127. Madsen et al., "A Population-Based Study of Measles, Mumps, and Rubella Vaccination and Autism"; Taylor et al., "Vaccines Are Not Associated with Autism: An Evidence-Based Meta-Analysis of Case-Control and Cohort Studies." See also discussion in Mnookin, *Panic Virus*.

128. Latour, *We Have Never Been Modern*; Latour, *Politics of Nature*; Shapin and Schaefer, *Leviathan and the Air-Pump*.

129. Pearce et al., "Beyond Counting Climate Consensus"; Oreskes and Cook, *Response to Pearce (In Press)*; Rice, "Beyond Climate Consensus."

130. The author of *Positively False: Exposing the Myths around HIV and AIDS*, Joan Shenton, became skeptical of modern medicine and Big Pharma after suffering severe iatragenic illness. From there she went on to denying the link between HIV and AIDS, and from there to become a climate change skeptic/denier. I have met Joan and once enjoyed a nice dinner with her at a conference. I do not doubt her experience of iatragenic illness, but I reject the slippery slope fallacy that has led her to suspect and reject science writ large.

131. "Pope Claims GMOs Could Have 'Ruinous Impact' on Environment."

132. Zycher, "Shut Up, She Explained."

133. My own work is the obvious reference here (Oreskes and Conway, *Merchants of Doubt*, and Supran and Oreskes, *Assessing ExxonMobil's Climate Change Communications*), but also the detailed documentation of industry obfuscation compiled by journalists, the Union of Concerned Scientists, and other NGOs: Banerjee, Song, and Hasemyer, "Exxon: The Road Not Taken"; Union of Concerned Scientists, "Exxon Mobil Report: Smoke Mirrors and Hot Air"; "Exxon Climate Denial Funding 1998–2014"; and The Royal Society, "Royal Society and Exxon Mobil."

134. Supran and Oreskes, *Assessing ExxonMobil's Climate Change Communications*.

135. Proctor and Schiebinger, *Agnotology*; Proctor, *Golden Holocaust*; Michaels, *Doubt Is Their Product*; Markowitz and Rosner, *Deceit and Denial*; Nestle, *Soda Politics*.

136. When this manuscript was in draft form, a reviewer raised the issue of classified scientific research, as well as unpublished work done inside industry. I take up the question of classified research in my forthcoming work, *Science on a Mission: American Oceanography from the Cold War to Climate Change*. I argue there that the secrecy that classification entailed did in fact have serious adverse consequences for oceanography. That said, a key element in classified research that might be used to

defend it as science is that much of it was in fact peer reviewed, albeit in classified journals. It may not have been *public* knowledge, but it was vetted by communities of experts and subject to critical interrogation. (For example, classified work in acoustics was not simply submitted to the US Navy, it was vetted by researchers in other institutions who had security clearances. This was, in my view, an imperfect system of vetting, but it was a system.) It meets the standard that I am advocating here. In contrast, a good deal of "in-house" industry research consists of reports that are not subject to open vetting. Indeed, this same reviewer noted that much tobacco industry science was "unpublished, of course," but "that does not mean it was not scientific research." That is a very interesting claim, but I would argue that if research is not published, then any vetting process is necessarily internal, and therefore not open. It would also be subject to the very conflicts of interest to which I am objecting. And consider this: when I worked in the mining industry, I wrote reports for my company on scientific topics, but they were not published. They do not appear on my CV. I am willing to say that those reports contained elements of scientific research, but they were not, in fact, science, because they were not subject to open vetting. And, perhaps justly, you will not find them in any science citation index or google scholar search.

137. Proctor and Schiebinger, *Agnotology*; Proctor, *Golden Holocaust*; Markowitz and Rosner, *Deceit and Denial*; Nestle, *Unsavory Truth*; Oreskes and Conway, *Merchants of Doubt*.

138. HADGirl, "10 Evil Vintage Cigarette Ads Promising Better Health." See also Brandt, *Cigarette Century*.

139. Oreskes and Conway, *Merchants of Doubt*; Michaels, *Doubt Is Their Product*.

Chapter 2. Science Awry

1. Wang, " 'Post-Truth' Named 2016 Word of the Year by Oxford Dictionaries."

2. Colbert, *"Post-Truth" Is Just a Rip-Off of "Truthiness."*

3. Jasanoff, *States of Knowledge*; Latour, *Science in Action*; see also Latour, *One More Turn after the Social Turn*.

Like any term that has come into widespread use, co-production is sometimes used in diverse ways—see http://scitechpopo.blogspot.com/2011/02/explaining-co-production.html. Jasanoff's student Clark Miller concludes that the key concept is that "the proposition that the ways in which we know and represent the world (both nature and society) are inseparable from the ways in which we choose to live in it" (Jasanoff, *States of Knowledge*, p. 2). On some level, this is an inarguable claim—no one can escape the world in which he or she lives. My concern is with the implication that all specific scientific knowledge claims are *themselves* co-produced. This would seem to imply that society at large plays an equal role in establishing them as do domain experts, and that there is no prospect, even in principle, of expert adjudication that stands apart from social context. While it seems to me inarguable that in practice experts can never stand wholly apart, it seems to me also inarguable that there are practices and ideals that expert groups follow in order to try to adjudicate knowledge claims on the basis of empirical adequacy, and in a manner that strives to be more rather than less independent of economic, religious, or other concerns. Objectivity, for example, is a regulative ideal that plays an important role in scientific evaluation of evidence. This does not mean it ever is, or could be, entirely achieved, but a community that holds onto it as a regulative ideal will likely produce different epistemic outcomes than one that does not.

4. On the relation between trust and social and cultural identity, see Wynne, *May the Sheep Safely Graze?*

5. Latour, Woolgar, and Salk, *Laboratory Life*, p. 285. One reviewer queries what Latour means by this. Of course, we would have to ask him, but I take this to mean that scientific claims are performances, to be accepted or rejected by their audiences.

6. Latour, "Has Critique Run Out of Steam?" Of course, the gist of most of Latour's writing is to problematize the distinction between matters of fact and matters of concern. More recently, Latour has acknowledged the cultural failure of climate scientists; see Latour, *Facing Gaia* and *Down to Earth*.

7. Latour, Woolgar, and Salk, *Laboratory Life*, p. 285. One reviewer queries what Latour means by this. Of course, we would have to ask him, but I take this to mean that scientific claims are performances, to be accepted or rejected by their audiences.

8. Leiserowitz and Smith, "Knowledge of Climate Change across Global Warming's Six Americas."

9. James, "Pragmatism's Conception of Truth," p. 222.

10. James, pp. 222–23.

11. Kuhn, *The Copernican Revolution*; Bloor, *The Enigma of the Aerofoil*.

12. Structural realists would argue that our prior theories, if they seemed to work, cannot have been wholly incorrect, but must have encapsulated some truth about the natural world, such as some element of its physical or mathematical structure. That is to say, there is some kind of continuity, either in form or structure even if not in content, between the old theory and the one that replaced it. See https://plato.stanford.edu/entries/structural-realism/.

13. On the role of the media, see also Ladher, *Nutrition Science in the Media*. On the misuse of statistics in nutrition, see Schoenfeld and Ioannides, "Is Everything We Eat Associated with Cancer?". On the adverse impact of industry disinformation see Lustig, *Fat Chance* and Nestle, *Soda Politics*.

14. Oreskes et al., "Viewpoint: Why Disclosure Matters." I believe one solution to corruption is fairly simple: evidence suggests many cases can be avoided through appropriate forms of self-awareness and disclosure, and those that are not avoided must be sanctioned. In many areas of science, there are few and generally only weak sanctions for violations of research integrity. For example, scientists are rarely sanctioned for failing to disclose external funding sources.

15. Cohen, *Revolution in Science*.

16. Oreskes, *The Rejection of Continental Drift*, introduction.

17. Laudan, "A Confutation of Convergent Realism." For qualifications, see Musgrave, "The Ultimate Argument for Scientific Realism." For another argument for convergent realism, see Hardin and Rosenberg, "In Defense of Convergent Realism." For a detailed summary and analysis of the arguments surrounding scientific realism, see Psillos, *Scientific Realism*.

18. Oreskes, *The Rejection of Continental Drift*.

19. In his popular book *Facing Up*, Nobel Laureate Steven Weinberg claimed that there are "truths that once discovered will form a permanent part of human knowledge." Weinberg, *Facing Up*, p. 201. One charitable view of this claim is to say that yes, there may be some such truths; the problem is that we have no way of knowing which ones they are! But I think the problem is deeper than this: because scientists typically do not study their own history, and because old knowledge can often be translated into the language of the new, scientists often are not aware or do not recognize the actual loss of knowledge over time. They assume that new work adds to and builds on the old—that science is cumulative—rather than seeing the ways in which old knowledge has been discarded or inadvertently lost. They do not recognize the two frontiers of knowledge: that which we are about to discover, and that which was discovered long ago and we are about to forget.

20. This reminds me of a well-known joke about an old man who was asked, "Have you lived in Vermont your whole life?" to which he replies, "Not yet."

21. This paragraph is taken from Oreskes, *The Rejection of Continental Drift*, p. 3.

22. See, for example, Feldman, "Climate Scientists Defend IPCC Peer Review as Most Rigorous in History."

23. This account is drawn, with her permission, from the Masters thesis of my former student, Katharine Saunders Bateman, "Sex in Education: A Case Study of the Establishment of Scientific Authority in the Service of a Social Agenda."

24. Showalter and Showalter, "Victorian Women and Menstruation," p. 86.

25. Further proof that old ideas never entirely die: Evidently president Donald Trump believes something similar to this: that each human being is somewhat like a non-rechargeable battery, containing a finite amount of energy. This apparently is why he chooses not to exercise. See Rettner, "Trump Thinks That Exercising Too Much Uses Up the Body's 'Finite' Energy."

26. The idea of "energeticism" became significant in late nineteenth-century biology, see, for example, William Coleman's classic work, *Biology in the Nineteenth Century*.

27. Bateman, "Sex in Education: A Case Study of the Establishment of Scientific Authority in the Service of a Social Agenda," p. 8.

28. Clarke, *Sex in Education; or, a Fair Chance for Girls*, p. 37.

29. Bateman, "Sex in Education: A Case Study of the Establishment of Scientific Authority in the Service of a Social Agenda," p. 9.

30. Ibid., p. 3.

31. In fairness, educators at the time did also warn against excessive sexual activity in men, particularly onanism. See Barker-Benfield, *The Culture of Sensibility: Sex and Society in Eighteenth-Century Britain*. But men could control their sexual activity through discipline; women could not control their reproductive systems.

32. Bateman, "Sex in Education: A Case Study of the Establishment of Scientific Authority in the Service of a Social Agenda," p. 4.

33. Clarke, *Sex in Education; or, a Fair Chance for Girls*, p. 140. Here we see as well the influence of Darwinian thinking.

34. Paul, "Eugenic Anxieties, Social Realities, and Political Choices," pp. 676–77; Kevles, *In the Name of Eugenics*, p. 111.

35. Bateman, "Sex in Education: A Case Study of the Establishment of Scientific Authority in the Service of a Social Agenda," p. 16.

36. Clarke, *Sex in Education; or, a Fair Chance for Girls*.

37. Bateman, "Sex in Education: A Case Study of the Establishment of Scientific Authority in the Service of a Social Agenda," p. 20.

38. Ibid., p. 23.

39. Ibid., p. 25.

40. Hall, *Adolescence*, p. 589. In 1874 the State Board of Health of Massachusetts did a survey of 160 physicians and school administrators, which was published in Popular Science Monthly, as well as in Clarke's 1874 sequel to *Sex in Education*, called *The Building of the Brain*. They asked respondents to answer, "based on personal observation," the question, "Is one sex more liable than the other to suffer from health from attendance in school?" One hundred and nine respondents said females were more liable than males, 1 said males more than females, 31 said equally liable, and 4 said neither. One hundred and twenty said puberty increases this liability. It is unclear how respondents were selected. See Bateman, "Sex in Education: A Case Study of the Establishment of Scientific Authority in the Service of a Social Agenda," p. 18. This was also sometimes cited as support for the Limited Energy Therapy, but note that this study provided no empirical evidence; it was essentially an opinion poll. This is one reason that I am personally dubious of expert elicitation as a form of evidence in contemporary science.

41. As Dorothy E. Roberts argued in her recent Tanner Lectures at Harvard, we can easily imagine similar claims being resurrected today, or in the future, in new genetic packages. Roberts, *The Ethics of Biosocial Science | The New Biosocial and the Future of Ethical Science*.

42. This discussion is drawn from my first book, Oreskes, *The Rejection of Continental Drift*.

43. Ibid., p. 65; Gould, *Ever since Darwin*, p. 161.

44. Oreskes, *The Rejection of Continental Drift*, p. 120.

45. Ibid., p. 126.

46. Ibid., p. 156.

47. Laudan, *From Mineralogy to Geology*.

48. Oreskes, *The Rejection of Continental Drift*, p. 136.

49. Ibid.

50. Hallam, *Great Geological Controversies*.

51. Chamberlin, "Investigation versus Propagandism."

52. Oreskes, *The Rejection of Continental Drift*, p. 139.

53. Ibid., p. 151.

54. Ibid., p. 227.

55. Ibid., chs. 5–6.

56. Ibid., pp. 192–96.

57. The most famous example is Michael Crichton, but his arguments have been often repeated. See for example, Crichton, "Why Politicized Science Is Dangerous." His book *State of Fear* was entirely based on the premise that climate science was the contemporary equivalent to eugenics. When Crichton first started making this claim, I was surprised that historians of eugenics were not stepping up to the plate to counter it. So I did, thus initiating the line of thinking that has led to this set of lectures. Oreskes, "Fear-Mongering Crichton Wrong on Science." See also Ekwurzel, "Crichton Thriller: State of Fear."

58. *Buck v. Bell*, 274 US. 4, 5, 9–10, 20, 76.

59. *Buck v. Bell*, 274 US.

60. Kevles, *In the Name of Eugenics*, p. 111.

61. Some would argue that "soft" eugenics has never really gone away, for example, when parents use abortion to terminate the pregnancies of children carrying genetic diseases. I am not sure this is eugenics; there is a world of difference in my view between individuals making free choices about reproduction and the government enforcing such choices, but I also recognize that those worlds can collide. See, for example Duster, *Backdoor to Eugenics*. Comfort, *The Science of Perfection*, argues that not only did eugenics not go away, but it became the heart of American medicine.

62. Laughlin's expertise is a bit hard to characterize. He was trained as a school teacher, and taught agriculture before moving to the Eugenics Record Office in 1910. He subsequently received a PhD in cytology, in 1917, and then, infamously, an honorary degree from the University of Heidelberg for his work on model sterilization laws, in 1936. See "Biography of Harry H. Laughlin."

63. Malthus, *An Essay on the Principle of Population, as It Affects the Future Improvement of Society. With Remarks on the Speculations of Mr. Godwin, M. Condorcet and Other Writers.*

64. Galton, *Hereditary Genius.*

65. Galton, *Memories of My Life*, p. 331.

66. Full disclosure: I am a member of the Save the Redwoods League.

67. "The Immigration Act of 1924 (The Johnson-Reed Act)."

68. Gould, *Bully for Brontosaurus*, 162.; On Hitler, see Spiro, *Defending the Master Race*; Kuhl, *The Nazi Connection.*

69. Grant and Osborn, *The Passing of the Great Race; or, the Racial Basis of European History. 4th Rev. Ed., with a Documentary Supplement, with Prefaces by Henry Fairfield Osborn*, p. 49.

70. Although there were Lamarckian versions of eugenics, as well.

71. In 1921, the Station for Experimental Evolution and the ERO were combined into the Carnegie Department of Genetics, with Charles Davenport as its director. The ERO function was terminated in 1939. See Allen, "The Eugenics Record Office"; Witkowski and Inglis, "Davenport's Dream," and Witkowski, *The Road to Discovery: A Short History of Cold Spring Harbor Laboratory.*

72. Davenport, *Eugenics, the Science of Human Improvement by Better Breeding*, p. 34.

73. The work was funded by Mrs. Mary Harriman, the widow of railroad magnate E. H. Harriman and mother of Averill, later governor of New York. Comfort, *The Tangled Field*, p. 79.

74. "The Immigration Act of 1924 (The Johnson-Reed Act)."

75. Historians have noted that Carrie's daughter was not in fact feeble-minded; she died of colitis at the age of eight after doing well in school under the care of a foster family. See Gould, "Carrie Buck's Daughter." In 1978, in the case of *Stump v. Sparkman*, the Supreme Court affirmed the right of a judge to approve a mother's request for the sterilization of a mildly mentally retarded fifteen-year-old girl. The girl later married; when she discovered she had been sterilized, she and her husband sued the state of Indiana. The case made it to the Supreme Court, who did not adjudicate the merits of the judge's action but the conclusion that he was immune from prosecution because he was acting in a judicial capacity. Opponents of the decision argued this was incorrect because he had failed to follow basic due process, such as appointing a guardian for the girl. White, *Stump v. Sparkman*, 435 U.S. 349.

76. Proctor, *Racial Hygiene*, p. 101.

77. Kevles, *In the Name of Eugenics*.

78. In this sense we may see eugenics and the Limited Energy Theory as related, which leads to the question: did feminists oppose eugenics? This is a topic that invites further scrutiny. Margaret Sanger's 1922 book, *The Pivot of Civilization*, appears to share many conventional eugenic attitudes of her time. See Sanger and Wells, *The Pivot of Civilization*. However, her biographer, Ellen Chesler, notes that Sanger has been often misquoted on eugenic matters, particularly by latter-day enemies of the organization she founded, Planned Parenthood. Chesler notes that Sanger opposed racial and ethnic stereotyping and "framed poverty as a matter of differential access to resources, including birth control, not as the immutable consequence of low inherent ability or intelligence or character." See Chesler, *Woman of Valor*, p. 484. Prominent members of the Catholic Church, including Pope Pius XI, also opposed eugenics, as an inappropriate intervention in God's plan, and argued that the source of degeneracy is not inheritance but sin. See Pope Pius XI, "Casti Connubii: Encyclical of Pope Pius XI on Christian Marriage to the Venerable Brethren, Patriarchs, Primates, Archbishops, Bishops, and Other Local Ordinaries Enjoying Peace and Communion with the Apostolic See."

79. Crichton, *State of Fear*; Dykes, "Late Author Michael Crichton Warned of Global Warming's Dangerous Parallels to Eugenics."

80. I say short answer because one could argue that the meanings of the word eugenics were multiple (as Kevles suggests) and changed over time, but even allowing for that, it is very clear that important social scientists and geneticists rejected the central claims of the eugenics movement, and did so quite explicitly.

81. Allen, *Eugenics and Modern Biology: Critiques of Eugenics, 1910–1945*.

82. Boas, "Eugenics," p. 471.

83. Boas, *Anthropology and Modern Life*.

84. Bowman-Kruhm, *Margaret Mead*, p. 140.

85. Paul argues that this was more true in Europe, particularly Scandinavia, where populations tended to be more homogenous than in the United States, but her own work, and Kevles's, shows that a great deal of attention in the United States involved immigrant populations that, by today's standards, would be considered "white," such as Irish and Italian immigrants. For a recent discussion of eugenics targeted at African Americans, see Roberts, *Killing the Black Body*.

86. This is not to say that only socialist geneticists objected to eugenics, but rather that socialists were conspicuous in their objections, which were tied to their politics. For an example of a non-socialist American geneticist who also argued for a better understanding of the interaction of inheritance and environment, see Jennings, "Heredity and Environment."

87. Oreskes, "Objectivity or Heroism?"

88. Kevles, *In the Name of Eugenics*, p. 167.

89. Ibid., p. 134.

90. Muller et al., "Social Biology and Population Improvement." That this question was being posed and in this way is revelatory about the state of the world at that time—not just in Germany—but the answer is revelatory as well.

91. In the end, every field is "special." Expertise does not travel well across disciplinary boundaries.

92. Muller et al., "Social Biology and Population Improvement." See also Darwin, "The Geneticists' Manifesto." The signatories included Huxley and Haldane.

93. Kevles, *In the Name of Eugenics*, p. 261.

94. Muller et al., "Social Biology and Population Improvement, " p. 521.

95. Ibid.

96. Ibid.

97. One of Muller's cosigners was the great Julian Huxley, one of the founders of the modern evolutionary synthesis—the successful joining of the Darwinian theory of selection with quantitative genetics. Like Muller, Huxley accepted the basic premise of eugenics, although he thought it more accurate to consider "ethnic" rather than "racial" groups as the appropriate unit of study. He also criticized negative eugenics such as involuntary sterilization. Yet he supported voluntary sterilization and birth control aimed at encouraging the "elimination of the few lowest and degenerate types," and in the 1960s, worried that many heralded humanitarian benefits of modern medicine and improvements in food supply were keeping alive individuals who would otherwise have died. This, he felt, clearly pushed against the natural tendencies that acted to improve the fitness of humans. While he was not clear on how it should be done, Huxley believed that humans needed to find a way to counter this dysgenic effect, and put genetic evolution back on track: As he put it: "we must manage to put it back on its age-old course of positive improvement." Kevles, *In the Name of Eugenics*, p. 261.

98. Allen, *Eugenics and Modern Biology*, p. 315. See also Spencer and Paul, "The Failure of a Scientific Critique," and *Did Eugenics Rest on an Elementary Mistake?*

99. Jennings, *The Biological Basis of Human Nature*. See also Jennings, "Heredity and Environment," *Scientific Monthly*, 1924. On Jennings's context, and other Americans who also criticized eugenics, see Allen, *Eugenics and Modern Biology: Critiques of Eugenics, 1910–1945*. Allen also discusses the interesting objections of American journalist Walter Lippmann.

100. Jennings, *Human Nature*, pp. 178–79.

101. Ibid.

102. Ibid., pp. 203–6.

103. Ibid., p. 206.

104. Ibid., p. 208.

105. Ibid., pp. 228–29.

106. Allen, *Eugenics and Modern Biology*.

107. Ibid., p.322.

108. See for example Paul, Spencer, et. al. *Eugenics at the Edges of Empire*.

109. Allen, *Eugenics and Modern Biology*, p. 324, makes an important additional point about this: much of the scientific critique was published in scientific journals or expressed privately (e.g., in letters) and therefore not known to the general public. He writes "In this way, the general public clearly got the impression that eugenics had the stamp of approval of the scientific/genetics community." He concludes that it is important, therefore, in contemporary debates, for experts to expose fallacies in public scientific claims, so the public can know of and understand the relevant critiques. I agree, and would add that it is also important for scientists to affirm in public the consensus on matters like vaccination safety when public discourse implies otherwise.

110. Skovlund et al., "Association of Hormonal Contraception with Depression."

111. Bakalar, "Contraceptives Tied to Depression Risk."

112. McDermott, "Can Birth Control Cause Depression?"

113. Tello, "Can Hormonal Birth Control Trigger Depression?" The issue of self-reporting is complex. On the role of patients in the AIDs crisis, see Epstein, *Impure Science*. My argument here is not that patients are necessarily correct in their judgments, but that patient reports are one form of evidence to which doctors and researchers would do well to attend.

114. A good starting point on the evidence surrounding autism and vaccines is Mnookin, *The Panic Virus*.

115. Tello, "Can Hormonal Birth Control Trigger Depression?"

116. For more about quantitative statistics in science, see Porter, *Trust in Numbers*; Daston and Galison, *Objectivity*.

117. On the prevalence of misdiagnosis, see Kirch and Schafii, "Misdiagnosis at a University Hospital in 4 Medical Eras"; On the effect of marketing on prescriptions see Iizuka and Jin, "The Effect of Prescription Drug Advertising on Doctor Visits."

118. Jones, Mosher, and Daniels, "Current Contraceptive Use in the United States, 2006–10, and Changes in Patterns of Use since 1995." The critical review can be found here: Schaffir, Worly, and Gur, "Combined Hormonal Contraception and Its Effects on Mood."

119. Christin-Maitre, "History of Oral Contraceptive Drugs and Their Use Worldwide."

120. Seaman and Dreyfus, *The Doctor's Case against the Pill*, p. 210.

121. Recently, a major study (again, in Denmark!) concluded that hormonal contraception use by women is associated with a doubled risk of suicide. Skovlund et al., "Association of Hormonal Contraception with Depression."

122. Thompson, "A Brief History of Birth Control in the U.S."

123. Seaman and Dreyfus, *The Doctor's Case against the Pill*, p. 213.

124. Ibid., 214.

125. Ibid., p. 223.

126. Oreskes, "The Scientific Consensus on Climate Change."

127. Cook et al., "Quantifying the Consensus"; Cook et al., "Consensus on Consensus."

128. Behre et al., "Efficacy and Safety of an Injectable Combination Hormonal Contraceptive for Men."

129. Ibid. The study concluded: "The study regimen led to near-complete and reversible suppression of spermatogenesis. The contraceptive efficacy was relatively good compared with other reversible methods available for men. The frequencies of mild to moderate mood disorders were relatively high." Media reports include: CNN, "Male Birth Control Shot Found Effective, but Side Effects Cut Study Short"; "Male Birth Control Study Killed after Men Report Side Effects"; Watkins, "Why the Male 'Pill' Is Still So Hard to Swallow."

130. Moses-Kolko et al., "Age, Sex, and Reproductive Hormone Effects on Brain Serotonin-1A and Serotonin-2A Receptor Binding in a Healthy Population"; Toufexis et al., "Stress and the Reproductive Axis."

131. Mayo Clinic Staff, "Antidepressants"; Oláh, "The Use of Fluoxetine (Prozac) in Premenstrual Syndrome"; Cherney and Watson, "Managing Antidepressant Sexual Side Effects."

132. Balon, "SSRI-Associated Sexual Dysfunction"; Rosen, Lane, and Menza, "Effects of SSRIs on Sexual Function."

133. Rosen, Lane, and Menza, "Effects of SSRIs on Sexual Function"; Block, "Antidepressant Killing Your Libido?"

134. Skovlund et al., "Association of Hormonal Contraception with Depression."

135. On this, see Ziliak and McCloskey, *Cult of Statistical Significance*, and Krosnick, this volume.

136. Oreskes and Conway, *Merchants of Doubt*, p. 141.

137. Bill Bechtel writes that scientists, particularly in the life sciences, often use mechanism to explain natural phenomena, but the role of mechanism in scientific methodologies has not been adequately explored in the philosophy of science, which has focused on deductive methods. Bechtel, *Discovering Cell Mechanisms*; Bechtel, *Mental Mechanisms*. See also the work of Carl Craver: Craver and Darden, *In Search of Mechanisms*; Machamer, Darden, and Craver, "Thinking about Mechanisms."

138. Saint Louis, "Feeling Guilty about Not Flossing?"

139. "Haven't Flossed Lately?"

140. Greenberg, "Science Says Flossing Doesn't Work. You're Welcome."

141. Clarke-Billings, "After Generations of Dentists' Advice, Has the Flossing Myth Been Shattered?"

142. Donn, "Medical Benefits of Dental Floss Unproven."

143. "Sink Your Teeth into This Debate over Flossing."

144. Donn, "Medical Benefits of Dental Floss Unproven." What none of these reports pointed out is all the other important uses of dental floss, including escaping from prison: "Inmate Recalls How He Flossed Way To Freedom." See also "Haven't Flossed Lately?"

145. "Everything You Believed about Flossing Is a Lie."

146. Rubin, "At a Loss over Dental Floss."

147. O'Connell, "The Great Dental Floss Scam."

148. Rubin, "At a Loss over Dental Floss."

149. *Take Care Staff*, "How a Journalist Debunked a Decades Old Health Tip."

150. Ibid.

151. Hare, "How an AP Reporter Took down Flossing."

152. Ghani, "The Deceit of the Dental Health Industry and Some Potent Alternatives." Nutrition certainly plays a role, as do genetics. But the question was whether or not flossing helps, so in this case these other factors are not at issue.

153. Donn, "Medical Benefits of Dental Floss Unproven."

154. Ibid.

155. Ibid. See also Resnick, "If You Don't Floss Daily, You Don't Need to Feel Guilty."

156. Harrar, "Should You Bother to Floss Your Teeth?"

157. Hare, "How an AP Reporter Took down Flossing."

158. "About Us | Cochrane."

159. Sambunjak et al., "Flossing for the Management of Periodontal Diseases and Dental Caries in Adults."

160. Ibid.

161. Although the effect size was small: "the SMD being −0.36 (95% CI −0.66 to −0.05) at 1 month, SMD −0.41 (95% CI −0.68 to −0.14) at 3 months and SMD −0.72 (95% CI −1.09 to −0.35) at 6 months."

162. The Silness-Löe index is an index of oral hygiene based on plaque accumulation. Moslehzadeh, "Silness-Löe Index." SMD refers to size effect.

163. Levine, "The Last Word on Flossing Is Two Words: Pascal's Wager."

164. Bakalar, "Gum Disease Tied to Cancer Risk in Older Women"; Smyth, "Gum Disease Sufferers 70% More Likely to Get Dementia." The comments following this article are notably hostile and negative; there appears to be a huge amount of resistance to accepting these findings. It would be worthwhile to investigate why readers of the *Times* are so hostile to the idea that flossing is so important!

165. Saint Louis, "Feeling Guilty about Not Flossing?" For more on dentists' defense of flossing, see Selleck, "National Media's Spotlight on Flossing Enables Dental Professionals to Shine."

166. Nancy Cartwright has argued that randomized controlled trials are not always the best tool for evaluating medical efficacy: Cartwright, "Are RCTs the Gold Standard?"; Cartwright, "A Philosopher's View of the Long Road from RCTs to Effectiveness."

167. Cartwright and Hardie, *Evidence-Based Policy*.

168. This point was made by Shmerling, "Tossing Flossing?"

169. Harrar, "Should You Bother to Floss Your Teeth?"

170. Rubin, "At a Loss over Dental Floss." See also "The Medical Benefit of Daily Flossing Called into Question."

171. Bechtel, *Mental Mechanisms*; Bechtel, *Discovering Cell Mechanisms*; Craver and Darden, *In Search of Mechanisms*; Machamer, Darden, and Craver, "Thinking about Mechanisms."

172. Saint Louis, "Feeling Guilty about Not Flossing?"

173. Cartwright and Hardie, *Evidence-Based Policy*.

174. "The Medical Benefit of Daily Flossing Called into Question."

175. Saint Louis, "Feeling Guilty about Not Flossing?"

176. Ibid.

177. Crichton, "Aliens Cause Global Warming"; Perry, "For Earth Day"; Bushway, "Eugenics," and many more.

178. "From the Editor."

179. Institute of Medicine (US) Committee on the Robert Wood Johnson Foundation Initiative on the Future of Nursing, "Transforming Leadership"; Editorial Board, "Opinion: Are Midwives Safer Than Doctors?". Also see Bourdieu and Passeron, *Reproduction in Education, Society and Culture*, for a discussion of how authoritative expertise is created and reproduced. We should not assume that "other professionals" are necessarily correct in their views simply because they are "on the ground"; many American farmers were enthusiastic proponents of eugenics, but then we could argue about whether these were really *experts* . . . and so on.

180. Finlayson, *Fishing for Truth*; On farmers, see the classic study by Brian Wynne, *May the Sheep Safely Graze?*, as well as his more recent work, e.g., "Participatory Mass Observation and Citizen Science." On patients as people who hold relevant knowledge, see Epstein, *Impure Science*.

181. Wynne, *May the Sheep Safely Graze?*, p. 61.

182. Garber, *Academic Instincts*.

183. Conversely, when scientists stray outside their domains of expertise and experience, we have a right to view their claims with caution, if not alarm. Elsewhere I have noted that a characteristic feature of "merchants of doubt" is their making

confident pronouncements that conflict with mainstream science, yet fall far outside their own domains of expertise. See discussion in chapter 1.

184. Wynne, *May the Sheep Safely Graze?*, p. 74 and again on p. 77.

185. Mnookin, *The Panic Virus*.

186. Oreskes and Conway, *Merchants of Doubt*; Brandt, *The Cigarette Century*; Proctor, *Golden Holocaust*; Proctor and Schiebinger, *Agnotology*; McGarity and Wagner, *Bending Science*; Wagner, "How Exxon Mobil 'Bends' Science to Cast Doubt on Climate Change"; Michaels, *Doubt Is Their Product*, 2008; Markowitz and Rosner, *Deceit and Denial*. A recent useful addition to this literature is Richard Staley, *Partisanal Knowledge: On Hayek and Heretics in Climate Science and Discourse*, in prep.

187. Mnookin, *The Panic Virus*.

188. https://www.nytimes.com/2015/01/04/opinion/sunday/playing-dumb-on-climate-change.html.

189. Bateman, "Sex in Education: A Case Study of the Establishment of Scientific Authority in the Service of a Social Agenda," p. 11.

190. For a similar argument in the world of technology, see Emily Chang's discussion of "meritocracy" in Silicon Valley in Chang, *Brotopia*.

191. Oreskes and Conway, *Merchants of Doubt*.

192. In his now-classic 1992 study of Cumbrian sheep farmers affected by radioactive fallout from the Chernobyl nuclear disaster, Brian Wynne concluded that lay people "showed themselves to be more ready than the scientific experts to reflect upon the status of their own knowledge" (Wynne, *Misunderstood Misunderstanding*, p. 298). His example may not be unique; many of us have found scientists often to be defensive when asked to reflect on the status of their knowledge, seeing any attempt to so reflect as an attack on that status, even when the request is made in the interest of making science stronger, or more objective (chapter 1). Wynne also raises another important point. He notes that "in the real world [*sic*] people have to reconcile or adapt to living with contradictions which are not necessarily within their control to dissolve," whereas "the implicit moral imperative driving science is to reorganize and control the world so as to iron out contradiction and ambiguity." No doubt he is right about this (although I would say that scientists live in the real world, too, even if they wish they did not), but it seems to me that increasingly scientists have come to recognize that, in many domains, they are unable to reorganize and control the world and must find ways to live with contradiction and ambiguity. Again, a feminist perspective may be useful here: women scientists spend their lives living with the contradiction of being women and scientists in a world that still tells them that the qualities of being female (read: emotional) are at odds with those of being scientific (read: rational). In this sense women scientists live with the sort of "double consciousness" of which W.E.B. DuBois famously wrote. See also Wynne's follow-up paper, *May the Sheep Safely Graze?*

193. Indeed, we would not now study it, had it not been. We would just view it as an idea that turned out not to be true . . .

194. Rudner, "The Scientist qua Scientist Makes Value Judgments"; Douglas, "Inductive Risk and Values in Science," 2000; Douglas, *Science, Policy, and the Value-Free Ideal*; Elliot and Richards, *Exploring Inductive Risk*.

195. Stern, *The Economics of Climate Change*; Stern, *Why Are We Waiting?*; Nordhaus, *The Climate Casino*.

196. Brandt, *The Cigarette Century*; Proctor, *Golden Holocaust*; Michaels, *Doubt Is Their Product*, 2008; Oreskes and Conway, *Merchants of Doubt*.

197. Pearce et al., "Beyond Counting Climate Consensus."

198. Wynne, "Misunderstood Misunderstanding," p. 301.

199. Stark, *Behind Closed Doors*, p. 10.

200. "Gender Diversity in Senior Positions and Firm Performance." Discussed in Emily Chang, *Brotopia*, p. 251.

201. Heather Douglas, *Science, Policy, and the Value-Free Ideal*.

202. This argument has sometimes been made to suggest that diversity does not matter in physics. Karen Barad has taken on this issue in her book *Meeting the Universe Halfway*, which begins with the wonderful line from a poem by Alice Fulton, "Because truths we don't expect have a hard time making themselves felt . . ."

203. Elsewhere I have argued that, had World War II not broken out, continental drift might well have been accepted in the 1940s, but whether or not that is true it is also relevant that when new information became available in the late 1950s and '60s, earth scientists embraced that information and rapidly developed plate tectonic theory. Oreskes, *Science on a Mission: American Oceanography in the Cold War and Beyond*.

204. Duesberg, "Peter Duesberg on AIDS."

205. Duesberg asserts, on his website, that his colleagues will not engage in debate, and uses this to solicit contributions to support his research. But his extensive publication record refutes this assertion. It may well be that he has trouble raising grant money for research in support of his hypothesis, but the funding environment is highly competitive, and it is reasonable that reviewers would not be supportive of spending money on investigations that they consider fruitless. Again, his colleagues may be wrong in their judgments, but this is not the same as suppressing dissent.

Coda. Values in Science

1. Zycher, "The Absurdity That Is the Paris Climate Agreement."

2. Philosopher Dale Jamieson has been eloquent on this point. See Jamieson, *Reason in a Dark Time*; and also Howe, *Behind the Curve*.

3. Heather Douglas makes this point, citing Herrick, "Junk Science and Environmental Policy," that the term "junk science" was used by critics to tarnish work whose implications they did not like (Douglas, *Science, Policy, and the Value-Free Ideal*, p. 11). What she misses, however, is that the term was not used promiscuously; it was promoted by *industry* groups wishing to tarnish science whose implications were adverse to their products or activities. Similarly, the term "sound science" was promulgated by the tobacco industry as a means to discredit science they wanted to tarnish as "unsound." Thus they created, with the help of PR men, TASSC: "The Advancement of Sound Science Coalition" (Oreskes and Conway, *Merchants of Doubt*, pp. 150–52).

4. Leiserowitz and Smith, "Knowledge of Climate Change across Global Warming's Six Americas"; see also Oreskes and Conway, *Merchants of Doubt*.

5. Moore, *Disrupting Science*, p. 23; Jewett, *Science, Democracy, and the American University*, particularly pp. 366–67.

6. Oreskes and Conway, *Merchants of Doubt*; Posner, *A Failure of Capitalism*; Brulle, "Institutionalizing Delay"; Dunlap and Brulle, *Climate Change and Society*; McCright and Dunlap, "Challenging Global Warming as a Social Problem"; McCright and Dunlap, "Social Movement Identity and Belief Systems."

7. Deen, "U.S. Lifestyle Is Not up for Negotiation"; "A Greener Bush."

8. Leiserowitz and Smith, "Knowledge of Climate Change across Global Warming's Six Americas."

9. Antonio and Brulle, "The Unbearable Lightness of Politics."

10. Numbers, *Galileo Goes to Jail and Other Myths about Science and Religion*, ch. 20.

11. K. Miller, *Only a Theory*, p. 139. See also Miller, *Finding Darwin's God*.

12. Proctor, *Value-Free Science?*; Douglas, *Science, Policy, and the Value-Free Ideal*.

13. Shapin, *A Social History of Truth*; Weber, "Science as a Vocation."

14. One example: Siegrist, Cvetkovich, and Gutscher, "Shared Values, Social Trust, and the Perception of Geographic Cancer Clusters." This paper argues that a lack of trust in public health experts is one reason why people believe in cancer clusters even when they have been shown to be statistical clusters and not causally related.

15. Merton, "Science and the Social Order," p. 329. See also discussion in Dant, *Knowledge, Ideology, Discourse*, and Mazotti, *Knowledge as Social Order*.

16. Daniels, "The Pure-Science Ideal and Democratic Culture"; Kevles, *The Physicists*; England, *A Patron for Pure Science. The National Science Foundation's Formative Years, 1945–57. NSF 82–24*; Greenberg, Maddox, and Shapin, *The Politics of Pure Science*.

17. Merton, "Science and the Social Order," p. 328.

18. Proctor, *Value-Free Science?*; Longino, *The Fate of Knowledge*; Douglas, *Science, Policy, and the Value-Free Ideal.*

19. Heilbron, *The Sun in the Church.*

20. Merton, *Science, Technology & Society in Seventeenth-Century England.*

21. Oreskes, *Science on a Mission.*

22. Oreskes and Krige, *Science and Technology in the Global Cold War;* Fleming, *Fixing the Sky;* Fleming, *Meteorology in America, 1800–1870;* Kohler, *Partners in Science;* H. S. Miller, *Dollars for Research.*

23. Unless those values are informed by falsehoods. In fact, I would argue this for a good deal of climate change denial. Right-wing commentators like Rush Limbaugh, Glenn Beck, and many members of the Cato Institute follow the arguments of Frederick von Hayek that social democracy inevitably leads toward totalitarianism. In the years since *The Road to Serfdom* was published, this claim has been shown to be untrue. See discussion in *Collapse of Western Civilization.* Similarly, many rural Americans dislike "big government" and the federal income tax because they think that the lion's share of government programs support people in the inner city. In fact, studies show that, per capita, rural America is the disproportionate recipient of federal funds. See Reeder and Bagi, "Federal Funding in Rural America Goes Far Beyond Agriculture," and Olson, "Study: Urban Tax Money Subsidizes Rural Counties."

24. In graduate school, I once asked my thesis advisor what pronoun I should use in writing a single-authored paper. He told me to avoid pronouns altogether and make the data the author: as in "the evidence suggests," "the data indicate," or the "results show"; to use the passive voice, "The ore deposit was produced by high-temperature fluids . . ."; or, when all else failed, to use the royal "we." Nowadays most scientific papers are coauthored, so the "I" problem is obviated. But these patterns help to explain why passive voice is so common in scientific writing.

25. Daston and Galison, *Objectivity.* This is related to the idea of the "view from nowhere." The authorless paper is the ideal expression of objective knowledge.

26. Correspondents pointed out the many biblical passages that note that only God knows when the end will come and it is presumptuous for ordinary men to suggest otherwise. See, for example, https://www.openbible.info/topics/when_the_world_will_end. Katherine Hayhoe points out that in 1 Thessalonians 4:9–12 and 2 Thessalonians 3:6–16, Paul addressed this exact point. "Many believe that some of the Thessalonians had stopped working because the end times were at hand. They might have felt that they were already living in God's kingdom, and there was no need to work; or they might have felt that Jesus was coming at any minute, and thus there was no point to work. The Thessalonian letters do speak quite a bit about misunderstandings about the end times, and it is interesting that the passages about idleness in 1 Thessalonians 4:9–12 and 2 Thessalonians 3:6–16 both come in the context of

teaching on the end times" (Hayhoe, email communication). Thus, even if the end is at hand, that is not justification for idleness or complacency. Another correspondent suggested to me that one might suggest to evangelical Christians that when Jesus arrives and finds the mess that we have made of his Father's work, he won't be happy! See also Mooney, "How to Convince Conservative Christians That Global Warming Is Real."

27. Dietz, "Bringing Values and Deliberation to Science Communication"; See also Fischhoff, "The Sciences of Science Communication"; National Academies of Sciences, *Using Science to Improve Science Communication*.

28. Heather Douglas makes a similar argument, although on slightly different grounds. She argues that the value-free ideal is not only unattainable (with which I agree) but undesirable (with which I agree but for different reasons). Her reason is that science as an activity should not be value-neutral, because it should not stand apart from society. For science to be an appropriate part of democratic society, scientists "must consider the consequences of their work as part of a basic responsibility we all share." From Douglas, *Science, Policy, and the Value-Free Ideal*, p. 15. I think that is probably correct, but my primary reason for rejecting value-neutrality is the argument here: that scientists-as individuals need to express and share their values to connect with our fellow citizens and build bonds of trust. My secondary reason is that we cannot be truly value-neutral (nor should we), so to claim that we are is to claim something that is impossible, which means either we are stupid, naïve, or lying, and that is no basis for building bonds of trust.

29. See for example, "Shaping Tomorrow's World: Our Values." For a practical example of values bridging the trust gap between scientists and science skeptics, see Webb and Hayhoe, "Assessing the Influence of an Educational Presentation on Climate Change Beliefs at an Evangelical Christian College."

30. Berlin, *Two Concepts of Liberty*; see also discussion in Baum and Nicols, *Isaiah Berlin and the Politics of Freedom*, p. 43. See also Abraham Lincoln: "We all declare for liberty, but in using the same word we do not all mean the same thing. . . . The shepherd drives the wolf from the sheep's throat, for which the sheep thanks the shepherd as a liberator, while the wolf denounces him for the same act as the destroyer of liberty."

31. Prothero, *Religious Literacy*.

32. Stern, *Report on the Economics of Climate Change*.

33. See my introduction on page ix of the Melville Press edition of Pope Francis, *Encyclical on Climate Change and Inequality*.

34. Intergovernmental Panel on Climate Change, *Global Warming of 1.5 °C*.

Chapter 3. The Epistemology of Frozen Peas: Innocence, Violence, and Everyday Trust in Twentieth-Century Science

1. For a particularly chilling account of the people and processes involved in current approaches to science, see Daniel Engber's detailed story about one-time Linus Pauling collaborator Art Robinson. Robinson has a scientific pedigree and a home laboratory. And he has played a key role in climate science skepticism. Engber, "The Grandfather of Alt-Science."

2. I recognize all the assumptions in the idea of an a priori relationship between science filtering "down" to become technology (or machinery), but I am using the assumptions because they are so widely shared. For a much more critical interrogation of the issue, see Pisano and Bussati, "Historical and Epistemological Reflections."

3. Several historians of technology, include Ruth Schwartz Cowan and Priscilla Brewer, have looked thoughtfully at the uses and meanings of domestic technologies. They have not, however, emphasized the high scientific origins of stoves or vacuum cleaners. Rather, historians of technology have been interested in how users modify and make sense of technologies and technological change. See also Jane Busch's fascinating 1983 paper about gas and electric cookers, though again, the focus is not on how scientific ideas mattered to the marketers and manufacturers of stoves.

4. Engdahl and Lidskog, "Risk Communication and Trust," have proposed that trust is not just rational/cognitive but also emotional, and current discussions on public trust "have a restricted rationalistic bias in which the cognitive-reflexive aspect of trust is emphasized at the expense of its emotional aspect," p. 704. Their work is an effort to develop a theory of trust that addresses this emotional character.

5. See for example Cartwright, *Hunting Causes and Using Them: Approaches in Philosophy and Economics.*

6. Remember Hofstadter, *Anti-Intellectualism in American Thought.*

7. Wynne, *Risk Management and Hazardous Wastes,* emphasizes that trust is relational and that the emotional–rational dichotomy is fundamentally misleading when analyzing public trust in science.

8. See Mitchell, *Test Cases: Reconfiguring American Law, Technoscience and Democracy in the Nuclear Pacific.*

9. Crawford, "Internationalism in Science."

10. On ideas of purity and new forms of "scientism," see Shapin, "The Virtue of Scientific Thinking."

11. http://www.vqronline.org/essay/technology-history-and-culture-appreciation-melvin-kranzberg.

12. Wang, "Physics, Emotion, and the Scientific Self."

13. July 9, 1954, Ernest Pollard to Thomas E. Murray, Commissioner to the Atomic Energy Commission, copy to Smyth, in Henry DeWolf Smyth Papers, American Philosophical Society, Philadelphia. Pollard was responding to Murray's recent statement that scientists doing any kind of defense work had to guard against all associations with those who could be suspicious. Pollard viewed this as "impossible" for those teaching in a university, who might have no means of discerning the loyalties of their students. Pollard said such an expectation could cause many university scientists to cease defense work.

14. July 9, 1954, Ernest Pollard to Thomas E. Murray, Commissioner to the Atomic Energy Commission, copy to Smyth, in Henry DeWolf Smyth Papers, American Philosophical Society, Philadelphia.

15. Galison, "Removing Knowledge."

16. Wang, *American Science in an Age of Anxiety*; Wang, "Physics, Emotion, and the Scientific Self."

17. See Steinberg below.

18. Wang, *American Science in an Age of Anxiety*; Moore, *Disrupting Science*; Bridger, *Scientists at War*.

19. Freire, "Science and Exile."

20. Smyth kept a file of public responses to his work and ideas. In Papers of Henry deWolf Smyth, American Philosophical Society, Philadelphia.

21. See particularly Steinberg's letter summarizing the accusations and rumors, December 11, 1953, Steinberg to Dr. Sydney Farber at the Children's Cancer Research Foundation, where he had been considered for a position, in Papers of Arthur Steinberg, American Philosophical Society. Steinberg was not offered the position because the dean had heard rumors of his Communist sympathies, though none of these rumors, he said, were correct.

22. See documents in Papers of Arthur Steinberg, American Philosophical Society Archives, Philadelphia. Scientists had in the course of the twentieth century even been imprisoned, like the geneticist Richard Goldschmidt in World War I, suspected of German sympathies, and like the cytogeneticist Masuo Kodani who did rubber research at the Japanese Internment Camp at Manzanar in World War II (Richmond, "A Scientist during Wartime"; Smocovitis, "Genetics behind Barbed Wire").

23. For a discussion of Luria's perspectives on the tensions of Cold War science, see Selya, *Salvador Luria's Unfinished Experiment*.

24. Probstein, "Reconversion and Non-Military Research Opportunities," 52.

25. Gusterson, *Testing Times*; Aaserud, "Sputnik and the 'Princeton Three'"; Cloud, "Imaging the World in a Barrel."

26. See forthcoming research by my PhD student Kathryn Dorsch.

27. Engber, "The Grandfather of Alt-Science."

28. Haraway, "Situated Knowledges," 598.

29. Haraway, "Situated Knowledges," 579.

30. Shapin, "What Else Is New?".

31. Edgerton, *The Shock of the Old Technology and Global History since 1900*.

32. For a compelling study of freezing things in science, one that has nothing to do with frozen peas, see Radin, *Life on Ice*.

Chapter 4. What Would Reasons for Trusting Science Be?

1. Hume, *A Treatise of Human Nature*, bk. 1, pt. 3, sec. 6; Hume, *An Enquiry Concerning Human Understanding*, secs. 4–5.

2. Descartes, *Meditations on First Philosophy*.

3. Sextus Empiricus, *Against the Logicians*, p. 179.

4. Sellars, "Empiricism and the Philosophy of Mind," sec. 38.

5. Lange, "Hume and the Problem of Induction."

6. Sellars, "Some Reflections on Language Games," p. 355; cf. Lange, "Would Direct Realism Resolve the Classical Problem of Induction?"

7. My thanks to an anonymous referee for kindly encouraging me to make this point more explicitly.

8. Kuhn, *The Structure of Scientific Revolutions*.

9. Galilei, *Two New Sciences*, p. 167.

10. Meli, "The Axiomatic Tradition in Seventeenth-Century Mechanics." In October 1643, Marin Mersenne mentioned these rival proposals in a letter to Theodore Deschamps. In his reply, Deschamps gave the same argument to show that neither Fabri's nor Le Cazre's proposal was correct. See Palmerino, "Infinite Degrees of Speed: Marin Mersenne and the Debate over Galileo's Law of Free Fall," pp. 295–96. I have followed Palmerino's elegant method of displaying the argument.

11. For more details on dimensional homogeneity, see Lange, *Because without Cause*, ch. 6 and the references given there.

12. Galileo's odd-number rule is not the only proposal regarding falling bodies that achieves dimensional homogeneity. Whereas the odd-number rule says that in the nth interval, the body covers $2n-1$ times the distance covered in the first interval, consider the rule that in the nth interval, the body covers $3n^2-3n+1$ times the distance covered in the first interval. On this proposal, the distances traversed in successive intervals are 1s, 7s, 19s, 37s, 61s, 91s, ... Like Galileo's rule, this proposal is dimensionally homogeneous. For example, in time intervals that are twice as long, the distances covered on this proposal are 8s ($= 1s + 7s$), 56s ($= 19s + 37s$), 152s ($= 61s + 91s$) ... —and the ratio of 8 to 56 to 152 is the ratio of 1 to 7 to 19. So although Galileo's dimensional argument rules out some rivals to the odd-number rule, it does not suffice to rule out every rival.

13. Newport, "In U.S., 46% Hold Creationist View of Human Origins."

14. Horowitz, "Paul Broun: Evolution, Big Bang 'Lies Straight From the Pit of Hell.'"

15. Hoffman, "Climate Science as Culture War."

16. Goodman, *Fact, Fiction, and Forecast*, pp. 59–83.

17. Laudan, "The Demise of the Demarcation Problem."

18. Laudan, "A Confutation of Convergent Realism."

Chapter 5. Pascal's Wager Reframed: Toward Trustworthy Climate Policy Assessments for Risk Societies

1. See https://www.epa.gov/sites/production/files/2017–10/documents/ria_proposed-cpp-repeal_2017–10.pdf, accessed November 30, 2017.

2. Beck, *Risk Society*, 21.

3. Kowarsch et al., "A Road Map for Global Environmental Assessments," showed that many decision-makers and stakeholders—having set ambitious mitigation goals in the Paris Agreement in 2015—want future assessments by the Intergovernmental Panel on Climate Change (IPCC) to focus more explicitly on the assessment of solution options, in particular policies, rather than problem analysis.

4. Sarewitz, "How Science Makes Environmental Controversies Worse."

5. Some mistakenly believe that the social sciences and humanities cannot, or should not (for moral reasons), provide objective assessments of policy pathways and their practical implications. This is argued for different reasons, including: (1) the theoretical assumption that it is impossible to deliberately steer social processes resulting in some form of general policy skepticism, (2) ethical relativism and radical constructivism, implying that any policy assessment is based on highly questionable, purely "subjective" value judgments, (3) the exclusive focus on politics and power structures in much of the science and technology studies in recent years, etc. We do not regard these as compelling reasons against engaging in scientific policy assessment to facilitate policy learning processes.

6. This wager was conceptualized by the French philosopher, mathematician, and physicist Blaise Pascal (1623–62) in a theological context.

7. IPCC, *Climate Change 2014*.

8. E.g., Koch et al., "Politics Matters."

9. Kowarsch et al., "A Road Map for Global Environmental Assessments."

10. Edenhofer and Kowarsch, "Cartography of Pathways."

11. Kowarsch, *A Pragmatist Orientation for the Social Sciences in Climate Policy*, ch. 5.

12. Kowarsch, *A Pragmatist Orientation for the Social Sciences in Climate Policy*, ch. 6.

13. Oreskes mentions the vivid example of astronomy promoted by the Catholic Church to calculate the exact Easter date with greater precision.

14. For more details see Putnam, *The Collapse of the Fact/Value Dichotomy and Other Essays* and Kowarsch 2016, sec. 6.2.3.

15. Edenhofer and Kowarsch "Cartography of Pathways"; Kowarsch, *A Pragmatist Orientation for the Social Sciences in Climate Policy*.

Chapter 6. Comments on the Present and Future of Science, Inspired by Naomi Oreskes

1. Bhattacharjee, "The Mind of a Con Man."

2. Bem, "Feeling the Future."

3. Yong, "A Failed Replication."

4. Lehrer, "The Truth Wears Off."

5. Vul et al., "Puzzlingly High Correlations."

6. Lehrer and Vul, "Voodoo Correlations."

7. Zimbardo, "The Stanford Prison Experiment."

8. Reicher and Haslam, "Rethinking the Psychology of Tyranny."

9. Festinger, *A Theory of Cognitive Dissonance*.

10. Lord, Ross, and Lepper, "Biased Assimilation and Attitude Polarization."

11. Miller et al., "The Attitude Polarization Phenomenon."

12. Carey, "Many Psychology Findings Not as Strong as Claimed."

13. LaCour and Green, "When Contact Changes Minds."

14. Carey and Belluck, "Doubts about Study of Gay Canvassers."

15. Mathews, "Papers in Economics 'Not Reproducible.'"

16. See, e.g., Barone, "Why Political Polls Are So Often Wrong."

17. Baker, "Biotech Giant Publishes Failures to Confirm High-Profile Science."

18. Open Science Collaboration, "Estimating the Reproducibility of Psychological Science."

19. Prinz, Schlange, and Asadullah, "Believe It or Not."

20. Walter, "Call to Arms on Data Integrity."

21. Koricheva and Gurevitch, "Uses and Misuses of Meta-analysis in Plant Ecology"; Jennions and Møller, "Relationships Fade with Time."

22. Freedman, Cockburn, and Simcoe, "The Economics of Reproducibility in Preclinical Research"; Freedman, "Lies, Damned Lies, and Medical Science."

23. Ioannidis, "Why Most Published Research Findings Are False."

24. John, Loewenstein, and Prelec, "Measuring the Prevalence of Questionable Research Practice."

Chapter 7. Reply

1. Oreskes, *Science on a Mission: American Oceanography from the Cold War to Climate Change*.

2. On the issue of retaining control of the intellectual agenda, see Forman, "Behind Quantum Electronics."

3. Bloor, *The Enigma of the Aerofoil*, gives several examples.

4. "Perceptions of Science in America."

5. Wazeck, *Einstein's Opponents*.

6. Oppenheimer et al., *Discerning Experts*; and Wolfe and Sharp, "Anti-Vaccinationists Past and Present."

7. Cook, Ellerton, and Kinkead, "Deconstructing Climate Misinformation to Identify Reasoning Errors"; Cook, Lewandowsky, and Ecker, "Neutralizing Misinformation through Inoculation"; Linden et al., "Inoculating against Misinformation"; Linden et al., "Inoculating the Public against Misinformation about Climate Change."

8. Layton, "Mirror-Image Twins."

9. Oreskes and Conway, *Merchants of Doubt*.

10. Oppenheimer et al., *Discerning Experts*.

11. Proctor, *Value-Free Science?*

12. Pope Francis, *Encyclical on Climate Change and Inequality*.

13. On the replication crisis in psychology see: Yong, "Psychology's Replication Crisis Is Running out of Excuses"; Bishop, "What Is the Reproducibility Crisis in Science and What Can We Do about It?"; "Oxford Reproducibility Lectures."

14. On the retraction crisis see Brainard and You, "What a Massive Database of Retracted Papers Reveals about Science Publishing's 'Death Penalty.'"

15. Gonzales and Cunningham, "The Promise of Pre-Registration in Psychological Research"; Nosek and Lindsay, "Preregistration Becoming the Norm in Psychological Science."

16. https://www.nature.com/articles/d41586-019-00857-9.

17. One particularly misguided response, in my view, was the 2018 proposal by a group of psychologists and other scientists to solve the replication problem by raising the threshold for statistical significance from 0.05 to 0.005: https://psyarxiv.com/mky9j/?_ga=2.29887741.370827084.1500902659-399963933.1500902659. One can only hope that these scientists read the 2019 *Nature* article.

18. Lewandowsky et al., "Seepage and Influence: An Evidence-Resistant Minority Can Affect Scientific Belief Formation and Public Opinion"; Lewandowsky et al., "The 'Pause' in Global Warming in Historical Context"; Lewandowsky et al., "Seepage"; Lewandowsky, Risbey, and Oreskes, "The 'Pause' in Global Warming";

Lewandowsky, Risbey, and Oreskes, "On the Definition and Identifiability of the Alleged 'Hiatus' in Global Warming"; Risbey et al., "A Fluctuation in Surface Temperature in Historical Context"; Risbey et al., "Well-Estimated Global Surface Warming in Climate Projections Selected for ENSO Phase."

19. Kennedy, "Why Did Earth's Surface Temperature Stop Rising in the Past Decade?"

20. Risbey et al., "A Fluctuation in Surface Temperature in Historical Context"; Mooney, "Ted Cruz Keeps Saying That Satellites Don't Show Global Warming. Here's the Problem"; Richardson, "Climate Change Whistleblower Alleges NOAA Manipulated Data to Hide Global Warming 'Pause' "; Taylor, "Global Warming Pause Extends Underwhelming Warming."

21. One study by Daniele Fanelli, "How Many Scientists Fabricate and Falsify Research?" found that out of more than eleven thousand scientists 2% (1.97, to be exact) had committed research fraud at least once in their career. I do not know how this number compares to doctors, lawyers, accountants, or investment advisors, but if it is correct, then 98% of scientists are thoroughly honest, which strikes me as a rather good figure. On the other hand, the same study reported that, "up to 33.7% admitted other questionable research practices." Clearly this invites further scrutiny of what constitutes questionable practices.

22. http://www.bbcprisonstudy.org/. It would be interesting to investigate the background to the BBC's sponsorship of this research.

23. One attempted replication concluded that an important factor involved self-selection by participants, and that small changes in the wording of the advertisement could affect outcomes. https://journals.sagepub.com/doi/abs/10.1177/0146167206292689?casa_token=6YVE-06G9BsAAAAA%3AwT8rDXdHa6jJp7vr qXo2bnPFOiCM5w7FFgrF26XsBlrJ7uJicqAlf3w3d3SLLxPWaeuyn-QMViuC. This also reminds us that psychology may be more vulnerable to the problem of non-replication of results, than, say physics, chemistry, or geology, because of how variable and suggestive human beings are. It would not be unreasonable to conclude, for example, that the Stanford experiment told us something interesting about how some people responded at a particular time to a particular set of circumstances, while acknowledging that small changes in those circumstances might yield different results.

24. Phillips, "The Female Mathematician Who Changed the Course of Physics— but Couldn't Get a Job."

25. Alberts et al., "Self-Correction in Science at Work."

26. This is an important reason why the tobacco industry long insisted that the science was not settled: if that had been true, then it might have been reasonable for the government to hold back on regulating tobacco. See Brandt, "Inventing Conflicts of Interest." On the other hand, if there was already significant evidence that tobacco

was likely harmful, then it might have well made sense for the government to begin to act to protect public health even in the absence of complete scientific agreement.

27. Hill, "The Environment and Disease: Association or Causation?"

28. For a more detailed exposition of this argument, drawing on practices in the oil and gas industry, see Oreskes, "Reconciling Representation with Reality."

29. Frederickson and Losada, " 'Positive Affect and the Complex Dynamics of Human Flourishing': Correction to Fredrickson and Losada (2005)."

30. Brown, Sokal, and Friedman, "The Complex Dynamics of Wishful Thinking"; Brown, Sokal, and Friedman, "The Persistence of Wishful Thinking."

31. For a full discussion, see Friedman and Brown, "Implications of Debunking the 'Critical Positivity Ratio' for Humanistic Psychology."

32. Steen, Casadevall, and Fang, "Why Has the Number of Scientific Retractions Increased?"

33. Fang, Steen, and Casadevall, "Misconduct Accounts for the Majority of Retracted Scientific Publications."

34. I am grateful to my colleague Alex Csiszar, who has studied the history of scientific publication, for his perspectives on this.

35. Steen, Casadevall, and Fang, "Why Has the Number of Scientific Retractions Increased?"

36. "Retraction Watch."

37. Retractions of journalistic announcements of findings in paleontology are fairly common, but they are generally cases of outright fraud (fake fossils) that are revealed by scientists after non-science media have reported unpublished reports. E.g., Pickrell, "How Fake Fossils Pervert Paleontology." One of the most famous hoaxes in the history of science occurred in paleontology—the Piltdown Man— reinforcing the idea that fraud and hoaxes may be more common in arenas that attract extensive public interest.

38. Siegel et al., "Methane Concentrations in Water Wells Unrelated to Proximity to Existing Oil and Gas Wells in Northeastern Pennsylvania."

39. Oreskes et al., "Viewpoint"; Tollefson, "Earth Science Wrestles with Conflict-of-Interest Policies."

40. Darrah et al., "The Evolution of Devonian Hydrocarbon Gases in Shallow Aquifers of the Northern Appalachian Basin."

41. I might be too generous here. There is lasting damage if industry-sponsored work keeps open a debate that would otherwise have appropriately closed. Of course, this is extremely difficult to judge, since there can be no "controlled experiment" in which the same data were collected and vetted absent industry involvement.

42. Myers et al., "Why Public Health Agencies Cannot Depend on Good Laboratory Practices as a Criterion for Selecting Data"; Saal et al., "Flawed Experimental

Design Reveals the Need for Guidelines Requiring Appropriate Positive Controls in Endocrine Disruption Research."

43. A controversial case is the paper by Gilles Seralini and colleagues of the effects of genetically modified maize and the Roundup herbicide on rats. The paper was retracted on the grounds that the findings were "inconclusive." Many scientists objected that this was inappropriate grounds for retraction; the paper was republished in a different journal. See Oransky, "Retracted Seralini GMO-Rat Study Republished." It was later revealed that the retraction had been orchestrated, or at least heavily influenced, by Monsanto, the manufacturer of Roundup. See McHenry, "The Monsanto Papers."

44. Brandt, *The Cigarette Century*; Proctor, *Golden Holocaust*; Michaels, *Doubt Is Their Product*, 2008; Oreskes and Conway, *Merchants of Doubt*. In 2013, the editors of the journals *BMJ, Heart, Thorax,* and *BMJ Open* stopped publishing research funded by the tobacco industry. In an editorial, they said that, "the tobacco industry, far from advancing knowledge, has used research to deliberately produce ignorance and to advance its ultimate goal of selling its deadly products while shoring up its damaged legitimacy." See Godlee et al., "Journal Policy on Research Funded by the Tobacco Industry."

45. Dugan, "In U.S., Smoking Rate Hits New Low at 16%." See also https://news .gallup.com/poll/237908/smoking-rate-hits-new-low.aspx.

46. Michaels, *Doubt Is Their Product*, 2008. See also Markowitz and Rosner, *Deceit and Denial*; Markowitz and Rosner, *Lead Wars*. Fanelli ("How Many Scientists Fabricate and Falsify Research?") finds that "misconduct was reported more frequently by medical/pharmacological researchers than others." This supports the possibility that misconduct in biomedicine is driven by the highly competitive atmosphere of medical research, the potentially distorting effects of interested funding, or both.

47. Michaels, *Doubt Is Their Product*, 2008; Michaels and Monforton, "Manufacturing Uncertainty"; Oreskes et al., "Viewpoint," July 7, 2015.

48. Franta and Supran, "The Fossil Fuel Industry's Invisible Colonization of Academia."

49. For a full discussion of facsimile science, see Oreskes, "Systematicity Is Necessary but Not Sufficient: On the Problem of Facsimile Science," *Synthèse*, https://link .springer.com/article/10.1007/s11229-017-1481-1.

50. Oberhaus, "Hundreds of Researchers from Top Universities Were Published in Fake Academic Journals."

51. On journals created or supported by the tobacco industry, see Proctor, *Golden Holocaust*.

52. Public Health Law Center, "United States v. Philip Morris (D.O.J. Lawsuit)"; Campaign for Tobacco-Free Kids, "Tobacco Companies Ordered to Place Statements about Products' Dangers on Websites and Cigarette Packs."

53. Oberhaus, "Hundreds of Researchers from Top Universities Were Published in Fake Academic Journals."

54. Carey, "A Peek Inside the Strange World of Fake Academia"; Wikipedia, "Predatory Conference."

55. Oberhaus, "Hundreds of Researchers from Top Universities Were Published in Fake Academic Journals." For the original study, in which scientists created a nonsense paper and submitted it to one of these journals, see https://www.daserste.de /information/reportage-dokumentation/dokus/videos/exclusiv-im-ersten-fake -science-die-luegenmacher-englische-version-video-100.html. Of course, nonsense can be published in legitimate journals as well, as demonstrated by the famous hoax by Alan Sokal and my discussion of the critical positivity ratio, above. But it is important to note that Sokal's hoax was perpetrated on a journal *Social Text,* which is not peer-reviewed. Some years ago, when Sokal was my upstairs neighbor, I asked him why he did not submit his hoax to a peer-reviewed journal, such as *Social Studies of Science*? (since his claim was that the field of science studies was largely nonsense). He replied, "Oh, well I knew the reviewers would see it was nonsense and it would be rejected." This actually reassured me: Sokal apparently did not think his hoax would get past peer review.

56. Open Science Collaboration, 2015. "Estimating the Reproducibility of Psychological Science," *Science.* 349: 943.

57. "Comment: Raise Standards for Preclinical Cancer Research." Begley, C. Glenn and Ellis, Lee M., 2012. *Nature* 483: 531–533.

58. A useful future line of investigation would be to examine how different social, intellectual, and political contexts produce different sorts of problems in different sciences. For example, in climate science we have documented conservatism—what my colleagues and I have called "erring on the side of least drama"—because the social pressure and intimidation has led climate scientist to be cautious (Brysse et al., 2013). In the oncology, and particularly in private sector research laboratories, there is an entirely different pressure: to move fast to be the first to prove a drug.

59. "Comment: Raise Standards for Preclinical Cancer Research." Begley, C. Glenn and Ellis, Lee M., 2012. *Nature* 483: 531–533, on p 532.

Afterword

1. For one perspective: Pomerantsev, "Why We're Post-Fact."

2. For a global perspective, see Ghosh, *The Great Derangement.*

3. Trump, "The Concept of Global Warming Was Created by and for the Chinese in Order to Make U.S. Manufacturing Non-Competitive." See also Jacobson, "Did Trump Say Climate Change Was a Chinese Hoax?" and Zurcher, "Does Trump Still

Think It's All a Hoax?" In casting doubt on the reality of climate change and disparaging climate science, Mr. Trump followed in the footsteps of many Republican politicians, including Oklahoma senator James Inhofe, who infamously sought to disprove climate change by bringing a snowball into the halls of Congress, and Texas senator Ted Cruz who repeatedly insisted that global warming had stopped, despite overwhelming scientific evidence to the contrary and various scientists attempts to set the record straight. (C-SPAN, *Sen. James Inhofe (R-OK) Snowball in the Senate.* Mooney, "Ted Cruz Keeps Saying That Satellites Don't Show Global Warming. Here's the Problem.") The claim that global warming had stopped was also picked up by numerous think tanks who have long promoted doubt about climate change and climate science, such as the Cato Institute: Bastasch and Maue, "Take a Look at the New 'Consensus' on Global Warming."

4. Smith, "Vaccine Rejection and Hesitancy."

5. "Where Is Glyphosate Banned?"; "IARC Monographs Volume 112: Evaluation of Five Organophosphate Insecticides and Herbicides."

6. Oppenheimer et al., *Discerning Experts.*

7. Rudwick, *The Great Devonian Controversy.*

8. Gross and Levitt, *Higher Superstition.*

9. Wang et al., "Recent Advances on Endocrine Disrupting Effects of UV Filters."

10. Downs et al., "Toxicopathological Effects of the Sunscreen UV Filter, Oxybenzone (Benzophenone-3), on Coral Planulae and Cultured Primary Cells and Its Environmental Contamination in Hawaii and the U.S. Virgin Islands." See also "Oxybenzone—Substance Information."

11. Gabbard et al., *Relating to Water Pollution.* The bill also bans the sale of octinoxate sunscreen, unless prescribed by a doctor. Octinoxate has also been implicated in coral toxicity. Schneider and Lim, "Review of Environmental Effects of Oxybenzone and Other Sunscreen Active Ingredients."

12. Jacobsen, "Is Sunshine the New Margarine?"

13. They also state that, "The U.S. Food & Drug Administration has approved the active ingredients in both types of sunscreen as safe and effective." The issue of FDA under-regulation, particularly of EDCs, is another matter. For a list of sunscreen ingredients allowed by the FDA see Code of Federal Regulations Title 21. For a discussion of questionable products allowed in US cosmetics but restricted or banned elsewhere see Becker, "10 American Beauty Ingredients That Are Banned in Other Countries." At present, oxybenzone is permitted in both the United States and in Europe up to 6% in sunscreens.

14. Perez, Musini, and Wright, "Effect of Early Treatment with Anti-Hypertensive Drugs on Short and Long-Term Mortality in Patients with an Acute Cardiovascular Event." Studies analyzing the effects of blood pressure on long-term cardiovascular health are recommended to take multiple samples so short-term variability does not

skew the data: "Age-Specific Relevance of Usual Blood Pressure to Vascular Mortality."

15. Consensus Development Panel, "National Institutes of Health Summary of the Consensus Development Conference on Sunlight, Ultraviolet Radiation, and the Skin. Bethesda, Maryland, May 8–10, 1989."

16. Cancer Council Australia, "Position Statement—Sun Exposure and Vitamin D—Risks and Benefits—National Cancer Control Policy."

17. Ibid. "In late autumn and winter in those parts of Australia where the UV Index is below 3, sun protection is not recommended. During these times, to support vitamin D production it is recommended that people are outdoors in the middle of the day with some skin uncovered on most days of the week. Being physically active while outdoors will further assist with vitamin D levels."

18. Cancer Council Australia, "SunSmart."

19. "Sunscreen Fact Sheet."

20. Ibid.

21. On the net risk of avoiding the sun: Lindqvist et al., "Avoidance of Sun Exposure as a Risk Factor for Major Causes of Death."

REFERENCES

Chapter 1

Agrawal, Arun. "Dismantling the Divide between Indigenous and Scientific Knowledge." *Development and Change* 26, no. 3 (July 1, 1995): 413–39. https://doi.org/10.1111/j.1467-7660.1995.tb00560.x.

Ayer, Alfred J. *Language, Truth and Logic*. 2nd edition. New York: Dover Publications, 1952.

Banerjee, Neela, Lisa Song, and David Hasemyer. "Exxon: The Road Not Taken." *InsideClimate News*, September 15, 2015. http://insideclimatenews.org/content/exxon-the-road-not-taken.

Barnes, Barry. *Interests and the Growth of Knowledge*. Routledge and Kegan Paul, 1977.

Berger, Peter L., and Thomas Luckmann. *The Social Construction of Reality: A Treatise in the Sociology of Knowledge*. New York: Anchor, 1967.

Berkman, Michael, and Eric Plutzer. *Evolution, Creationism, and the Battle to Control America's Classrooms*. 1st edition. New York: Cambridge University Press, 2010.

Bernard, Claude. *An Introduction to the Study of Experimental Medicine*. Translated by H. C. Greene. USA: Schuman, 1865. http://archive.org/details/b21270557.

Bloor, David. *Knowledge and Social Imagery*. Chicago: University of Chicago Press, 1991.

———. *The Enigma of the Aerofoil: Rival Theories in Aerodynamics, 1909–1930*. Chicago: University of Chicago Press, 2011.

Bourdeau, Michel. "Auguste Comte." In *The Stanford Encyclopedia of Philosophy*, edited by Edward N. Zalta, Winter 2015. Metaphysics Research Lab, Stanford University, 2015. https://plato.stanford.edu/archives/win2015/entries/comte/.

Campbell, Charles. "The Great Global Warming Hustle." baltimoresun.com. Accessed August 24, 2017. http://www.baltimoresun.com/news/opinion/oped/bs-ed-op-0721-global-warming-hoax-20170719-story.html.

Comte, Auguste. *Introduction to Positive Philosophy*. Indianapolis: Hackett Publishing, 1988.

Conant, James Bryant. *Harvard Case Histories in Experimental Science Volume I*. Harvard University Press, 1957. http://archive.org/details/harvardcasehisto 010924mbp.

Conis, Elena. "Jenny McCarthy's New War on Science: Vaccines, Autism and the Media's Shame." *Salon*, November 8, 2014. http://www.salon.com/2014/11/08 /jenny_mccarthys_new_war_on_science_vaccines_autism_and_the_medias _shame/.

Cook, John, et al. "Quantifying the Consensus on Anthropogenic Global Warming in the Scientific Literature." *Environmental Research Letters* 8 (024024). 2013.

Cook, John, et al. "Consensus on Consensus: A Synthesis of Consensus Estimates on Human-Caused Global Warming." *Environmental Research Letters* 11 (048002). 2016.

Coyne, Jerry. "Another Philosopher Proclaims a Nonexistent 'Crisis' in Evolutionary Biology." *Why Evolution Is True* (blog), 2012. https://whyevolutionistrue.wordpress .com/2012/09/07/another-philosopher-proclaims-a-nonexistent-crisis-in -evolutionary-biology/.

Crosland, Maurice. *Science under Control: The French Academy of Sciences 1795–1914*. Cambridge: Cambridge University Press, 2002.

Dant, Tim. *Knowledge, Ideology, and Discourse: A Sociological Perspective*. New York: Routledge, 2012. First edition, 1991.

Duhem, Pierre Maurice Marie. *The Aim and Structure of Physical Theory*. Translated by Philip P. Wiener. Reprint edition. Princeton, NJ: Princeton University Press, 1991.

Ellis, J., I. Mulligan, J. Rowe, and D. L. Sackett. "Inpatient General Medicine Is Evidence Based. A-Team, Nuffield Department of Clinical Medicine." *Lancet (London, England)* 346, no. 8972 (August 12, 1995): 407–10.

Epstein, Steven. *Impure Science: AIDS, Activism, and the Politics of Knowledge*. 1st edition. Berkeley: University of California Press, 1996.

Ernst, Edzard. "The Efficacy of Herbal Medicine—an Overview." *Fundamental & Clinical Pharmacology* 19, no. 4 (August 1, 2005): 405–9. https://www.ncbi.nim.nih .gov/pubmed/16011726.

"Evolution Resources from the National Academies." Accessed August 24, 2017. http://www.nas.edu/evolution/Statements.html.

"Exxon Climate Denial Funding 1998–2014." *Exxon Secrets*. Accessed October 11, 2018. https://exxonsecrets.org/html/index.php.

Fausto-Sterling, Anne. *Myths of Gender: Biological Theories about Women and Men, Revised Edition*. 2nd edition. New York: Basic Books, 1992.

Feyerabend, Paul. *Against Method*. London: Verso, 1993.

Fleck, Ludwik. "Scientific Observation and Perception in General." In *Cognition and Fact*, 59–78. Boston Studies in the Philosophy of Science. Dordrecht: Springer, 1986. https://doi.org/10.1007/978-94-009-4498-5_4.

Fleck, Ludwik, and Thomas S. Kuhn. *Genesis and Development of a Scientific Fact.* Edited by Thaddeus J. Trenn and Robert K. Merton. Translated by Frederick Bradley. Chicago: University of Chicago Press, 1981.

Friedman, Michael, and Richard Creath, eds. *The Cambridge Companion to Carnap.* Cambridge: Cambridge University Press, 2008.

Frodeman, Robert, and Adam Briggle. "When Philosophy Lost Its Way." *New York Times: Opinionator,* 2016. https://opinionator.blogs.nytimes.com/2016/01/11/when-philosophy-lost-its-way/.

Fuller, Steve. *Thomas Kuhn: A Philosophical History for Our Times.* Chicago: University of Chicago Press, 2000. http://www.press.uchicago.edu/ucp/books/book/chicago/T/bo3629340.html.

Galison, Peter. "History, Philosophy, and the Central Metaphor." *Science in Context* 2, no. 1 (1988).

Galison, Peter, and David J. Stump, eds. *The Disunity of Science: Boundaries, Contexts, and Power.* 1st edition. Stanford: Stanford University Press, 1996.

Giddens, Anthony. *The Consequences of Modernity.* 1st edition. Stanford: Stanford University Press, 1991.

Goonatilake, Susantha. *Toward a Global Science: Mining Civilizational Knowledge.* Bloomington: Indiana University Press, 1998.

Gross, Paul R., and Norman Levitt. *Higher Superstition: The Academic Left and Its Quarrels with Science.* Reprint edition. Baltimore: Johns Hopkins University Press, 1997.

Gross, Paul R., Norman Levitt, and Martin W. Lewis, eds. *The Flight from Science and Reason.* Baltimore: New York Academy of Sciences, 1997.

Hacking, Ian. *The Social Construction of What?* Revised edition. Cambridge, MA: Harvard University Press, 2000.

HADGirl. "10 Evil Vintage Cigarette Ads Promising Better Health." *Healthcare Administration Degree Programs* (blog). Accessed October 11, 2018. https://www.healthcare-administration-degree.net/10-evil-vintage-cigarette-ads-promising-better-health/.

Harding, Sandra. *The Science Question in Feminism.* 1st edition. Ithaca: Cornell University Press, 1986.

———. Women at the Center: History of Women's Studies at the University of Delaware. Video, July 20, 2012. MSS 664. University of Delaware women's studies oral history collection. http://udspace.udel.edu/bitstream/handle/19716/12708/Tape%20Log%20Sandra%20Harding.pdf.

Hayward, Steven F. "Climategate (Part II)." *American Enterprise Institute,* 2011. http://www.aei.org/publication/climategate-part-ii/.

Hemmer, Nicole. *Messengers of the Right: Conservative Media and the Transformation of American Politics.* Philadelphia: University of Pennsylvania Press, 2016.

Hicks, Stephen. "Is Newton's *Principia* a Rape Manual?" *Stephen Hicks, PhD* (blog), June 24, 2017. https://www.stephenhicks.org/2017/06/24/newtons-principia-as -a-rape-manual/.

Hubbard, Ruth. *The Politics of Women's Biology*. New Brunswick, NJ: Rutgers University Press, 1990.

Jones, Alex. "About Alex Jones." *Infowars*. Accessed August 15, 2017. https://www .infowars.com/about-alex-jones/.

Keller, Evelyn Fox. *Reflections on Gender and Science*. New Haven, CT: Yale University Press, 1995.

Kuhn, Thomas. "Reflections on My Critics." In *Criticism and the Growth of Knowledge: Volume 4: Proceedings of the International Colloquium in the Philosophy of Science, London, 1965*, by Imre Lakatos (ed.) and Alan Musgrave. Cambridge: Cambridge University Press, 1970.

Kuhn, Thomas S., and James Bryant Conant. *The Copernican Revolution: Planetary Astronomy in the Development of Western Thought*. Revised edition. Cambridge, MA: Harvard University Press, 1992.

Ladyman, James, Don Ross, David Spurrett, and John Collier. *Every Thing Must Go: Metaphysics Naturalized*. 1st edition. Oxford: Oxford University Press, 2009.

Lakatos, Imre. "Criticism and the Methodology of Scientific Research Programmes." *Proceedings of the Aristotelian Society*, New Series, 69 (1968): 149–86.

Laland, Kevin. "What Use Is an Extended Evolutionary Synthesis?" Presented at the International Society for History, Philosophy, and Social Studies of Science, Sao Paolo, Brazil, July 2017.

Laland, Kevin, Tobias Uller, Marc Feldman, Kim Sterelny, Gerd B. Müller, Armin Moczek, Eva Jablonka, et al. "Does Evolutionary Theory Need a Rethink?" *Nature News* 514, no. 7521 (October 9, 2014): 161. https://doi.org/10.1038/514161a.

Laland, Kevin N., Tobias Uller, Marcus W. Feldman, Kim Sterelny, Gerd B. Müller, Armin Moczek, Eva Jablonka, and John Odling-Smee. "The Extended Evolutionary Synthesis: Its Structure, Assumptions and Predictions." *Proc. R. Soc. B* 282, no. 1813 (August 22, 2015). https://doi.org/10.1098/rspb.2015.1019.

Latour, Bruno. *Science in Action: How to Follow Scientists and Engineers through Society*. Cambridge, MA: Harvard University Press, 1987.

———. *We Have Never Been Modern*. Translated by Catherine Porter. Cambridge, MA: Harvard University Press, 1993.

———. *Politics of Nature: How to Bring the Sciences into Democracy*. Translated by Catherine Porter. Cambridge, MA: Harvard University Press, 2004.

Latour, Bruno. *Facing Gaia: Eight Lectures on the New Climatic Regime*. Cambridge: Polity Press, 2017.

Longino, Helen E. *Science as Social Knowledge: Values and Objectivity in Scientific Inquiry*. Princeton, NJ: Princeton University Press, 1990.

———. *The Fate of Knowledge.* Princeton, NJ: Princeton University Press, 2001.

Lowery, Ilana. "Why Gender Diversity on Corporate Boards Is Good for Business." *Phoenix Business Journal,* November 27, 2017. https://www.bizjournals.com /phoenix/news/2017/11/27/why-gender-diversity-on-corporate-boards-is-good .html.

Madsen, Kreesten Meldgaard, Anders Hviid, Mogens Vestergaard, Diana Schendel, Jan Wohlfahrt, Poul Thorsen, Jørn Olsen, and Mads Melbye. "A Population-Based Study of Measles, Mumps, and Rubella Vaccination and Autism." *New England Journal of Medicine* 347, no. 19 (November 7, 2002): 1477–82. https://doi.org/10 .1056/NEJMoa021134.

Markowitz, Gerald, and David Rosner. *Deceit and Denial: The Deadly Politics of Industrial Pollution.* 1st paperback printing edition. Berkeley: University of California Press, 2003.

Michaels, David. *Doubt Is Their Product: How Industry's Assault on Science Threatens Your Health.* 1st edition. Oxford: Oxford University Press, 2008.

Miller, Kenneth R. *Only a Theory: Evolution and the Battle for America's Soul.* Reprint edition. New York: Penguin Books, 2009.

Mirowski, Philip, and Dieter Plehwe, eds. *The Road from Mont Pelerin: The Making of the Neoliberal Thought Collective.* 1st edition. Cambridge, MA: Harvard University Press, 2009.

Mnookin, Seth. *The Panic Virus: The True Story behind the Vaccine-Autism Controversy.* 1st edition. New York: Simon and Schuster, 2012.

Mößner, Nicola. "Thought Styles and Paradigms—a Comparative Study of Ludwik Fleck and Thomas S. Kuhn." *Studies in History and Philosophy of Science Part A, Model-Based Representation in Scientific Practice,* 42, no. 2 (June 1, 2011): 362–71. https://doi.org/10.1016/j.shpsa.2010.12.002.

Mohan, Kamlesh. *Science and Technology in Colonial India.* Delhi: Aakar Books, 2014.

Morris, William Edward, and Charlotte R. Brown. "David Hume." In *The Stanford Encyclopedia of Philosophy,* edited by Edward N. Zalta, Spring 2017. Metaphysics Research Lab, Stanford University, 2017. https://plato.stanford.edu/archives /spr2017/entries/hume/.

Motterlini, Matteo. *For and Against Method.* Chicago: University of Chicago Press, 1999.

National Center for Science Education. "Background on Tennessee's 21st Century Monkey Law." Accessed August 15, 2017. https://ncse.com/library-resource /background-tennessees-21st-century-monkey-law.

Nestle, Marion. *Unsavory Truth: How Food Companies Skew the Science of What We Eat.* New York: Basic Books, 2018.

Nestle, Marion, Mark Bittman, and Neal Baer. *Soda Politics: Taking on Big Soda.* 1st edition. Oxford: Oxford University Press, 2015.

Newport, Frank. "In U.S., 46% Hold Creationist View of Human Origins." Gallup.com, 2012. http://www.gallup.com/poll/155003/Hold-Creationist-View-Human-Origins.aspx.

Oppenheimer, Michael, Dale Jamieson, Naomi Oreskes, Keynyn Brysse, Jessica O'Reilly, Matthew Shindell, and Milena Wazeck. *Discerning Experts: The Practices of Scientific Assessment for Public Policy*. University of Chicago Press, 2019.

Oreskes, Naomi. *The Rejection of Continental Drift: Theory and Method in American Earth Science*. 1st edition. New York: Oxford University Press, 1999.

———. "The Devil Is in the (Historical) Details: Continental Drift as a Case of Normatively Appropriate Consensus?" [Essay Review of Miriam Solomon: *Social Epistemology*], *Perspectives in Science* 16, no. 2 (2008): 253–64.

———. "Why We Should Trust Scientists." *TED Talk*, 2014. https://www.ted.com/talks/naomi_oreskes_why_we_should_believe_in_science.

———. Response by Oreskes to "Beyond Counting Climate Consensus," *Environmental Communication* 11, no. 6 (2017): 731–37.

Oreskes, Naomi, Daniel Carlat, Michael E. Mann, Paul D. Thacker, and Frederick S. vom Saal. "Viewpoint: Why Disclosure Matters." *Environmental Science & Technology* 49, no. 13 (July 7, 2015): 7527–28. https://doi.org/10.1021/acs.est.5b02726.

Oreskes, Naomi, and Erik M. Conway. *Merchants of Doubt: How a Handful of Scientists Obscured the Truth on Issues from Tobacco Smoke to Global Warming*. Reprint edition. New York: Bloomsbury Press, 2011.

Oreskes, Naomi, Kristin Shrader-Frechette, and Kenneth Belitz. "Verification, Validation, and Confirmation of Numerical Models in the Earth Sciences." *Science* 263, no. 5147 (February 4, 1994): 641–46. https://doi.org/10.1126/science.263.5147.641.

Page, Scott E., and Katherine Phillips. *The Diversity Bonus: How Great Teams Pay Off in the Knowledge Economy*. Edited by Earl Lewis and Nancy Cantor. Princeton, NJ: Princeton University Press, 2017.

Pearce, Warren, Reiner Grundmann, Mike Hulme, Sujatha Raman, Eleanor Hadley Kershaw, and Judith Tsouvalis. "Beyond Counting Climate Consensus." *Environmental Communication* 11, no. 6 (July 23, 2017): 1–8. https://doi.org/10.1080/17524032.2017.1333965.

"Pope Claims GMOs Could Have 'Ruinous Impact' on Environment." *Genetic Literacy Project* (blog), 2016. https://geneticliteracyproject.org/2016/10/20/pope-claims-gmos-ruinous-impact-environment/.

Popper, Karl. *Conjectures and Refutations: The Growth of Scientific Knowledge*. New York: Basic Books, 1962.

———. *The Myth of the Framework: In Defence of Science and Rationality*. Edited by M. A. Notturno. 1st edition. London: Routledge, 1996.

Proctor, Robert N. *Golden Holocaust: Origins of the Cigarette Catastrophe and the Case for Abolition*. 1st edition. Berkeley: University of California Press, 2012.

Proctor, Robert N., and Londa Schiebinger, eds. *Agnotology: The Making and Unmaking of Ignorance*. 1st edition. Stanford: Stanford University Press, 2008.

Quine, Willard V. O. "Two Dogmas of Empiricism." *Philosophical Review* 60, no. 1 (1951): 20–43.

Quine, W. V., and Rudolf Carnap. *Dear Carnap, Dear Van: The Quine-Carnap Correspondence and Related Work: Edited and with an Introduction by Richard Creath*. Edited by Richard Creath. 1st printing edition. Berkeley: University of California Press, 1991.

Redd, Nola Taylor. "Wernher von Braun, Rocket Pioneer." Space.com. Accessed October 11, 2018. https://www.space.com/20122-wernher-von-braun.html.

Reisch, George. "Anticommunism, the Unity of Science Movement and Kuhn's Structure of Scientific Revolutions." *Social Epistemology* 17, no. 2–3 (January 1, 2003): 271–75. https://doi.org/10.1080/0269172032000144289.

Rice, Ken. "Beyond Climate Consensus." *AndThenThere'sPhysics*, July 30, 2017. https://andthentheresphysics.wordpress.com/2017/07/30/beyond-climate-consensus/.

Richards, Jay. "When to Doubt a Scientific 'Consensus.'" *American Enterprise Institute*, 2010. https://www.aei.org/publication/when-to-doubt-a-scientific-consensus/.

Richardson, Alan, and Thomas Uebel. *The Cambridge Companion to Logical Empiricism*. Cambridge: Cambridge University Press, 2007.

Rossiter, Margaret W. *Women Scientists in America: Struggles and Strategies to 1940*. JHU Press, 1984.

The Royal Society. "Royal Society and ExxonMobil." Accessed October 11, 2018. https://royalsociety.org/topics-policy/publications/2006/royal-society-exxonmobil/.

Sachs, Jeffrey. "How the AEI Distorts the Climate Debate." *Huffington Post* (blog), 2014. http://www.huffingtonpost.com/jeffrey-sachs/how-the-aei-distorts-the_b_4751680.html.

Sady, Wojciech. "Ludwik Fleck." In *The Stanford Encyclopedia of Philosophy*, edited by Edward N. Zalta, Summer 2016. Metaphysics Research Lab, Stanford University, 2016. https://plato.stanford.edu/archives/sum2016/entries/fleck/.

Sample, Ian. "Scientists Offered Cash to Dispute Climate Study." *Guardian*, 2007, sec. Environment. http://www.theguardian.com/environment/2007/feb/02/frontpagenews.climatechange.

Saxon, Wolfgang. "William B. Shockley, 79, Creator of Transistor and Theory on Race." *New York Times*, 1989. http://www.nytimes.com/learning/general/onthisday/bday/0213.html?mcubz=0.

Schiebinger, Londa. "Has Feminism Changed Science?" *Signs* 25, no. 4 (2000): 1171–75. https://www.jstor.org/stable/3175507.

Schiebinger, Londa, and Claudia Swan, eds. *Colonial Botany: Science, Commerce, and Politics in the Early Modern World.* Philadelphia: University of Pennsylvania Press, 2007.

Schreiber, Ronnee. *Righting Feminism: Conservative Women and American Politics.* 1st edition. Oxford: Oxford University Press, 2008.

Scott, Colin. "Science for the West, Myth for the Rest?" In *The Postcolonial Science and Technology Studies Reader,* edited by Sandra G. Harding, 175. Durham, NC: Duke University Press, 2011.

Semali, Ladislaus M., and Joe L. Kincheloe. *What Is Indigenous Knowledge?: Voices from the Academy.* New York: Routledge, 2002.

Shapin, Steven. *A Social History of Truth: Civility and Science in Seventeenth-Century England.* 1st edition. Chicago: University of Chicago Press, 1995.

Shapin, Steven, and Simon Schaefer. *Leviathan and the Air-Pump: Hobbes, Boyle, and the Experimental Life.* Princeton, NJ: Princeton University Press, 1985. http://www.jstor.org/stable/j.ctt7sv46.

Shenton, Joan. *Positively False: Exposing the Myths around HIV and AIDS.* London: I. B. Tauris, 1998.

Sokal, Alan. *Beyond the Hoax: Science, Philosophy and Culture.* 1st edition. Oxford: Oxford University Press, 2010.

Solomon, Miriam. *Social Empiricism.* Cambridge, MA: A Bradford Book, 2007.

Staley, Richard. "Partisanal Knowledge: On Hayek and Heretics in Climate Science and Discourse." Presented at the Weak Knowledge: Forms, Functions, and Dynamics, Frankfurt, July 4, 2017. http://www.hsozkult.de/event/id/termine-34489.

Stark, Laura. *Behind Closed Doors: IRBs and the Making of Ethical Research.* 1st edition. Chicago: University of Chicago Press, 2012.

Sterman, John D. "The Meaning of Models." *Science* 264, no. 5157 (April 15, 1994): 329–30. https://doi.org/10.1126/science.264.5157.329-b.

Supran, Geoffrey, and Naomi Oreskes. "Assessing ExxonMobil's Climate Change Communications (1977–2014)." *Environmental Research Letters* 12 (August 1, 2017): 084019. https://doi.org/10.1088/1748-9326/aa815f.

Taylor, Luke E., Amy L. Swerdfeger, and Guy D. Eslick. "Vaccines Are Not Associated with Autism: An Evidence-Based Meta-Analysis of Case-Control and Cohort Studies." *Vaccine* 32, no. 29 (June 17, 2014): 3623–29. https://doi.org/10.1016/j.vaccine.2014.04.085.

Union of Concerned Scientists. "Global Warming Skeptic Organizations." Accessed August 16, 2017. http://www.ucsusa.org/global_warming/solutions/fight-misinformation/global-warming-skeptic.html#.WZSl4P_yvL-.

———. "ExxonMobil Report: Smoke Mirrors & Hot Air." *Union of Concerned Scientists.* Accessed October 11, 2018. https://www.ucsusa.org/global-warming/solutions/fight-misinformation/exxonmobil-report-smoke.html.

Von Neumann, John. "Can We Survive Technology?" *Fortune*, 1955.

Walker, M. "Navigating Oceans and Cultures: Polynesian and European Navigation Systems in the Late Eighteenth Century." *Journal of the Royal Society of New Zealand* 42, no. 2 (June 1, 2012): 93–98. https://doi.org/10.1080/03036758.2012.673494.

Weinberg, Steven. *Facing Up: Science and Its Cultural Adversaries*. New edition. Cambridge, MA: Harvard University Press, 2003.

Weir, Todd H. *Secularism and Religion in Nineteenth-Century Germany: The Rise of the Fourth Confession*. Cambridge: Cambridge University Press, 2014.

Yearley, Steven, David Mercer, Andy Pitman, Naomi Oreskes, and Erik Conway. "Perspectives on Global Warming." *Metascience* 21, no. 3 (2012): 531–59.

Zammito, John H. *A Nice Derangement of Epistemes: Post-Positivism in the Study of Science from Quine to Latour*. 1st edition. Chicago: University of Chicago Press, 2004.

Zycher, Benjamin. "The Enforcement of Climate Orthodoxy and the Response to the Asness-Brown Paper on the Temperature Record." *American Enterprise Institute*, 2015. http://www.aei.org/publication/the-enforcement-of-climate-orthodoxy-and-the-response-to-the-asness-brown-paper-on-the-temperature-record/.

———. "Shut Up, She Explained: My Request for Climate Evidence." *American Enterprise Institute*, 2016. https://www.aei.org/publication/shut-up-she-explained-my-request-for-climate-evidence/.

Chapter 2

"About Us | Cochrane." Accessed August 27, 2017. https://us.cochrane.org/about-us.

Allen, Garland E. "The Eugenics Record Office at Cold Spring Harbor, 1910–1940: An Essay in Institutional History." *Osiris* 2 (1986): 225–64. https://www.jstor.org/stable/301835.

———. "Eugenics and Modern Biology: Critiques of Eugenics, 1910–1945." *Annals of Human Genetics* 75, no. 3 (May 2011): 314–25. https://doi.org/10.1111/j.1469-1809.2011.00649.x.

Bakalar, Nicholas. "Contraceptives Tied to Depression Risk." *New York Times*, September 30, 2016, sec. Wellness. https://www.nytimes.com/2016/09/30/well/live/contraceptives-tied-to-depression-risk.html.

———. "Gum Disease Tied to Cancer Risk in Older Women." *New York Times*, August 2, 2017. https://www.nytimes.com/2017/08/02/well/gum-disease-tied-to-cancer-risk-in-older-women.html.

Balon, Richard. "SSRI-Associated Sexual Dysfunction." *American Journal of Psychiatry* 163, no. 9 (September 1, 2006): 1504–9. https://doi.org/10.1176/ajp.2006.163.9.1504.

Barad, Karen. *Meeting the Universe Halfway: Quantum Physics and the Entanglement of Matter and Meaning*. Second printing edition. Durham, NC: Duke University Press, 2007.

Barker-Benfield, G. J. *The Culture of Sensibility: Sex and Society in Eighteenth-Century Britain*. Chicago: University of Chicago Press, 1992. http://www.press.uchicago.edu/ucp/books/book/chicago/C/bo3625409.html.

Bateman, Katharine Saunders. "Sex in Education: A Case Study of the Establishment of Scientific Authority in the Service of a Social Agenda." Masters of Arts in Liberal Studies, Dartmouth College, 1994.

Bechtel, William. *Mental Mechanisms: Philosophical Perspectives on Cognitive Neuroscience*. 1st edition. New York: Psychology Press, 2007.

———. *Discovering Cell Mechanisms: The Creation of Modern Cell Biology*. 1st edition. Cambridge: Cambridge University Press, 2008.

Behre, Hermann M., Michael Zitzmann, Richard A. Anderson, David J. Handelsman, Silvia W. Lestari, Robert I. McLachlan, M. Cristina Meriggiola, et al. "Efficacy and Safety of an Injectable Combination Hormonal Contraceptive for Men." *Journal of Clinical Endocrinology & Metabolism* 101, no. 12 (December 1, 2016): 4779–88. https://doi.org/10.1210/jc.2016–2141.

"Biography of Harry H. Laughlin." Accessed October 13, 2018. http://library.truman.edu/manuscripts/laughlinbio.asp.

Block, Jenny. "Antidepressant Killing Your Libido? Not for Long." *Fox News*, October 11, 2011. http://www.foxnews.com/health/2011/10/10/antidepressant-killing-your-libido-not-for-long.html.

Bloor, David. *The Enigma of the Aerofoil: Rival Theories in Aerodynamics, 1909–1930*. Chicago: University of Chicago Press, 2011.

Boas, Franz. "Eugenics." *Scientific Monthly* 3, no. July–December (1916): 471–78. http://www.estherlederberg.com/Franz_Boaz.pdf.

———. *Anthropology and Modern Life*. New York: Norton, 1962. http://archive.org/details/anthropologymodeooboas.

Bourdieu, Pierre, and Jean-Claude Passeron. *Reproduction in Education, Society and Culture*. Thousand Oaks, CA: SAGE, 1977.

Bowman-Kruhm, Mary. *Margaret Mead: A Biography*. Greenwood Publishing Group, 2003.

Brandt, Allan. *The Cigarette Century: The Rise, Fall, and Deadly Persistence of the Product That Defined America*. 1st reprint edition. New York: Basic Books, 2009.

Bushway, Rob. "Eugenics: When Scientific Consensus Leads to Mass Murder." *Climate Depot*, April 10, 2017. http://www.climatedepot.com/2017/04/10/eugenics-when-scientific-consensus-leads-to-mass-murder/.

Cartwright, Nancy. "Are RCTs the Gold Standard?" *BioSocieties* 2, no. 1 (March 1, 2007): 11–20. https://doi.org/10.1017/S1745855207005029.

———. "A Philosopher's View of the Long Road from RCTs to Effectiveness." *Lancet* 377, no. 9775 (April 23, 2011): 1400–1401. https://doi.org/10.1016/S0140–6736(11)60563–1.

Cartwright, Nancy, and Jeremy Hardie. *Evidence-Based Policy: A Practical Guide to Doing It Better.* 1st edition. Oxford: Oxford University Press, 2012.

Chamberlin, T. C. "Investigation versus Propagandism." *Journal of Geology* 27, no. 5 (1919): 305–38. https://doi.org/10.2307/30059365.

Chang, Emily. *Brotopia: Breaking Up the Boys' Club of Silicon Valley.* New York: Portfolio, 2018.

Cherney, Kristeen, and Kathryn Watson. "Managing Antidepressant Sexual Side Effects." *Healthline,* March 3, 2016. http://www.healthline.com/health/erectile-dysfunction/antidepressant-sexual-side-effects.

Chesler, Ellen. *Woman of Valor: Margaret Sanger and the Birth Control Movement in America.* New York: Simon and Schuster, 2007.

Christin-Maitre, Sophie. "History of Oral Contraceptive Drugs and Their Use Worldwide." *Best Practice & Research. Clinical Endocrinology & Metabolism* 27, no. 1 (February 2013): 3–12. https://doi.org/10.1016/j.beem.2012.11.004.

Clarke, Edward H. *Sex in Education; or, a Fair Chance for Girls.* Houghton, Mifflin, and Company, 1873.

Clarke-Billings, Lucy. "After Generations of Dentists' Advice, Has the Flossing Myth Been Shattered?" *Newsweek,* August 3, 2016. http://www.newsweek.com/after-generations-recommendation-has-flossing-myth-been-shattered-486761.

CNN, Susan Scutti. "Male Birth Control Shot Found Effective, but Side Effects Cut Study Short." *CNN.* Accessed August 27, 2017. http://www.cnn.com/2016/10/30/health/male-birth-control/index.html.

Cohen, I. Bernard. *Revolution in Science.* Cambridge, MA: The Belknap Press of Harvard University Press, 1985.

Colbert, Stephen. *"Post-Truth" Is Just a Rip-Off of "Truthiness." The Late Show with Stephen Colbert,* 2016. https://www.youtube.com/watch?v=CkoyqUoBY7M.

Coleman, William. *Biology in the Nineteenth Century: Problems of Form, Function and Transformation.* 2nd edition. Cambridge: Cambridge University Press, 1978.

Comfort, Nathaniel. *The Tangled Field: Barbara McClintock's Search for the Patterns of Genetic Control.* Cambridge, MA: Harvard University Press, 2009.

———. *The Science of Human Perfection: How Genes Became the Heart of American Medicine.* Reprint edition. New Haven, CT: Yale University Press, 2014.

Craver, Carl F., and Lindley Darden. *In Search of Mechanisms: Discoveries across the Life Sciences.* Chicago: University of Chicago Press, 2013.

Crichton, Michael. "Aliens Cause Global Warming: A Caltech Lecture by Michael Crichton." Michelin Lecture, Caltech, January 17, 2003. https://wattsupwiththat.com/2010/07/09/aliens-cause-global-warming-a-caltech-lecture-by-michael-crichton/.

———. *State of Fear.* Reprint edition. New York: Harper, 2009.

Crichton, Michael. "Why Politicized Science Is Dangerous." *MichaelCrichton.Com* (blog). Accessed October 13, 2018. http://www.michaelcrichton.com/why -politicized-science-is-dangerous/.

Dant, Tim. *Knowledge, Ideology, and Discourse: A Sociological Perspective.* New York: Routledge, 2012. First edition, 1991.

Darwin, Leonard. "The Geneticists' Manifesto." *Eugenics Review* 31, no. 4 (January 1940): 229–30. http://www.ncbi.nlm.nih.gov/pmc/articles/PMC2962351/.

Daston, Lorraine J., and Peter Galison. *Objectivity.* New York: Zone Books, 2010.

Davenport, Charles Benedict. *Eugenics, the Science of Human Improvement by Better Breeding.* New York: H. Holt and Company, 1910. http://archive.org/details /eugenicsscienceo00daverich.

Donn, Jeff. "Medical Benefits of Dental Floss Unproven." *AP News,* 2016. https:// apnews.com/f7e66079d9ba4b4985d7af350619a9e3/medical-benefits-dental-floss -unproven.

Douglas, Heather. *Science, Policy, and the Value-Free Ideal.* Pittsburgh: University of Pittsburgh Press, 2009.

Duesberg, Peter. "Peter Duesberg on AIDS." *Duesberg on AIDS.* Accessed August 27, 2017. http://www.duesberg.com/.

Duster, Troy. *Backdoor to Eugenics.* 2nd edition. New York: Routledge, 2003.

Dykes, Aaron. "Late Author Michael Crichton Warned of Global Warming's Dangerous Parallels to Eugenics." *Infowars* (blog), 2009. https://www.infowars.com /late-author-michael-crichton-warned-of-global-warmings-dangerous-parallels -to-eugenics/.

The Editorial Board. "Opinion: Are Midwives Safer Than Doctors?" *New York Times,* December 14, 2014, sec. Opinion. https://www.nytimes.com/2014/12/15/opinion /are-midwives-safer-than-doctors.html.

Ekwurzel, Brenda. "Crichton Thriller State of Fear." *Union of Concerned Scientists,* 2005. https://www.ucsusa.org/global-warming/solutions/fight-misinformation /crichton-thriller-state-of.html.

Elliot, Kevin, and Ted Richards. *Exploring Inductive Risk: Case Studies of Values in Science.* Oxford: Oxford University Press, 2017.

"Everything You Believed about Flossing Is a Lie." TheWeek.com. August 2, 2016. http://theweek.com/speedreads/640513/everything-believed-about -flossing-lie.

Feldman, Stacy. "Climate Scientists Defend IPCC Peer Review as Most Rigorous in History." *InsideClimate News,* February 26, 2010. http://insideclimatenews.org/news /20100226/climate-scientists-defend-ipcc-peer-review-most-rigorous-history.

Finlayson, Alan Christopher. *Fishing for Truth: A Sociological Analysis of Northern Cod Stock Assessments from 1977 to 1990.* St. John's, NL: Institute of Social & Economic, 1994.

"From the Editor." *Hedgehog Review*. Accessed August 27, 2017. http://www.iasc -culture.org/THR/THR_article_2016_Fall_Editor.php.

Hemment, Drew, Rebecca Ellis, and Brian Wynne. "Participatory Mass Observation and Citizen Science." *Leonardo* 44, no. 1 (February 2011): 62–63. https://doi.org /10.1162/LEON_a_00096.

Galton, Francis. *Hereditary Genius: An Inquiry into Its Laws and Consequences*. New York: Macmillan, 1869.

———. *Memories of My Life*. London: Methuen and Company, 1908. http://archive .org/details/memoriesmylife01galtgoog.

Garber, Marjorie. *Academic Instincts*. Princeton, NJ: Princeton University Press, 2003.

"Gender Diversity in Senior Positions and Firm Performance: Evidence from Europe." *IMF*. Accessed October 13, 2018. https://www.imf.org/en/Publications/WP /Issues/2016/12/31/Gender-Diversity-in-Senior-Positions-and-Firm-Performance -Evidence-from-Europe-43771.

Ghani, Maseeh. "The Deceit of the Dental Health Industry and Some Potent Alternatives." *Collective Evolution*, 2016. http://www.collective-evolution.com/2016/10 /26/the-deceit-of-the-dental-health-industry-and-some-potent-alternatives/.

Gould, Stephen Jay. "Carrie Buck's Daughter." *Constitutional Commentary* 2 (1985): 331–40.

———. *Bully for Brontosaurus: Reflections in Natural History*. Reprint edition. New York: W. W. Norton and Company, 1992.

———. *Ever since Darwin: Reflections in Natural History*. New York: W. W. Norton and Company, 1992.

Grant, Madison, and Henry Fairfield Osborn. *The Passing of the Great Race; or, the Racial Basis of European History. 4th Rev. Ed., with a Documentary Supplement, with Prefaces by Henry Fairfield Osborn*. New York: Scribner, 1922. http://archive.org /details/passingofgreatra00granuoft.

Greenberg, Will. "Science Says Flossing Doesn't Work. You're Welcome." *Mother Jones* (blog). Accessed August 27, 2017. http://www.motherjones.com/environment /2016/08/flossing-doesnt-work/.

Hall, Granville Stanley. *Adolescence*. New York: D. Appleton and Company, 1904. http://archive.org/details/adolescenceitsp01hallgoog.

Hallam, A. *Great Geological Controversies*. 2nd edition. Oxford: Oxford University Press, 1989.

Hardin, Clyde L., and Alexander Rosenberg. "In Defense of Convergent Realism." *Philosophy of Science* 49, no. 4 (December 1, 1982): 604–15. https://doi.org/10.1086 /289080.

Hare, Kristen. "How an AP Reporter Took down Flossing." *Poynter*, August 4, 2016. http://www.poynter.org/2016/how-an-ap-reporter-took-down-flossing/424625/.

Harrar, Sari. "Should You Bother to Floss Your Teeth?" *Consumer Reports*. Accessed August 27, 2017. http://www.consumerreports.org/beauty-personal-care/should -you-bother-to-floss-your-teeth/.

"Haven't Flossed Lately? Don't Feel Too Bad: Evidence for the Benefits of Flossing Is 'Weak, Very Unreliable.'" *Los Angeles Times*, August 2, 2016. http://www.latimes .com/science/sciencenow/la-sci-floss-benefits-unproven-20160802-snap-story .html.

Iizuka, Toshiaki, and Ginger Zhe Jin. "The Effect of Prescription Drug Advertising on Doctor Visits." *Journal of Economics & Management Strategy* 14, no. 3 (September 1, 2005): 701–27. https://doi.org/10.1111/j.1530-9134.2005.00079.x.

"The Immigration Act of 1924 (The Johnson-Reed Act)." Office of the Historian. Accessed August 27, 2017. https://history.state.gov/milestones/1921–1936 /immigration-act.

"Inmate Recalls How He Flossed Way to Freedom." DeseretNews.com, August 14, 1994. http://www.deseretnews.com/article/369688/INMATE-RECALLS-HOW -HE-FLOSSED-WAY-TO-FREEDOM.html.

Institute of Medicine (US) Committee on the Robert Wood Johnson Foundation Initiative on the Future of Nursing. "Transforming Leadership." In *The Future of Nursing: Leading Change, Advancing Health*. National Academies Press (US), 2011. https://www.ncbi.nlm.nih.gov/books/NBK209867/.

James, William. "Pragmatism's Conception of Truth." *Journal of Philosophy, Psychology and Scientific Methods* 4, no. 6 (1907): 141–55. https://doi.org/10.2307 /2012189.

Jennings, Herbert Spencer. "Heredity and Environment." *Scientific Monthly* 19, no. 3 (1924): 225–38. https://www.jstor.org/stable/7321.

———. *The Biological Basis of Human Nature*. 1st edition. New York: W. W. Norton and Company, 1930.

Jones, Jo, William Mosher, and Kimberly Daniels. "Current Contraceptive Use in the United States, 2006–10, and Changes in Patterns of Use since 1995." National Health Statistics Report. National Center for Health Statistics, October 18, 2012. https:// www.cdc.gov/nchs/data/nhsr/nhsr060.pdf.

Kevles, Daniel J. *In the Name of Eugenics: Genetics and the Uses of Human Heredity*. Cambridge, MA: Harvard University Press, 1985.

Kirch, W., and C. Schafii. "Misdiagnosis at a University Hospital in 4 Medical Eras." *Medicine* 75, no. 1 (January 1996): 29–40. https://doi.org/10.1097/00005792 -199601000-00004.

Kuhl, Stefan. *The Nazi Connection: Eugenics, American Racism, and German National Socialism*. New York: Oxford University Press, 2002.

Ladher, Navjoyt. "Nutrition Science in the Media: You Are What You Read." *BMJ* 353 (April 7, 2016): i1879. https://doi.org/10.1136/bmj.i1879.

Latour, Bruno. "One More Turn after the Social Turn: Easing Science Studies into the Non-Modern World," 1992, 25.

———. "Why Has Critique Run Out of Steam? From Matters of Fact to Matters of Concern." *Critical Inquiry* 30, no. 2 (January 1, 2004): 225–48. https://doi.org/10.1086/421123.

Latour, Bruno, Steve Woolgar, and Jonas Salk. *Laboratory Life: The Construction of Scientific Facts.* 2nd edition. Princeton, NJ: Princeton University Press, 1986.

Laudan, Larry. "A Confutation of Convergent Realism." *Philosophy of Science* 48, no. 1 (1981): 19–49. https://doi.org/10.2307/187066.

Laudan, Rachel. *From Mineralogy to Geology: The Foundations of a Science, 1650–1830.* Chicago: University of Chicago Press, 1987.

Leiserowitz, Anthony, and Nicholas Smith. "Knowledge of Climate Change across Global Warming's Six Americas." Yale Project on Climate Change Communication. New Haven, CT: Yale University, 2010. http://climatecommunication.yale.edu/publications/knowledge-of-climate-change-across-global-warmings-six-americas/.

Levine, Timothy. "The Last Word on Flossing Is Two Words: Pascal's Wager." *Chicago Tribune: Digital Edition.* Accessed August 27, 2017. http://digitaledition.chicagotribune.com/tribune/article_popover.aspx?guid=e22b8ba6-f7c1-43ee-a10a-6af692511143.

Lustig, Robert H. *Fat Chance: The Bitter Truth about Sugar.* Fourth Estate, 2013.

Machamer, Peter, Lindley Darden, and Carl F. Craver. "Thinking about Mechanisms." *Philosophy of Science* 67, no. 1 (March 1, 2000): 1–25. https://doi.org/10.1086/392759.

"Male Birth Control Study Killed after Men Report Side Effects." NPR.org. Accessed August 27, 2017. http://www.npr.org/sections/health-shots/2016/11/03/500549503/male-birth-control-study-killed-after-men-complain-about-side-effects.

Malthus, T. R. (Thomas Robert). *An Essay on the Principle of Population, as It Affects the Future Improvement of Society. With Remarks on the Speculations of Mr. Godwin, M. Condorcet and Other Writers.* London: J. Johnson, 1798. http://archive.org/details/essayonprinciploomalt.

Markowitz, Gerald, and David Rosner. *Deceit and Denial: The Deadly Politics of Industrial Pollution.* 1st paperback printing edition. Berkeley: University of California Press, 2003.

Mayo Clinic Staff. "Antidepressants: Get Tips to Cope with Side Effects." *Mayo Clinic.* Accessed August 27, 2017. http://www.mayoclinic.org/diseases-conditions/depression/in-depth/antidepressants/art-20049305.

Mazotti, Massimo. *Knowledge as Social Order: Rethinking the Sociology of Barry Barnes.* New York: Routledge, 2016.

McDermott, Annette. "Can Birth Control Cause Depression?" *Healthline*, 2016. http://www.healthline.com/health/birth-control/birth-control-and-depression.

McGarity, Thomas O., and Wendy E. Wagner. *Bending Science: How Special Interests Corrupt Public Health Research*. Cambridge, MA: Harvard University Press, 2012.

"The Medical Benefit of Daily Flossing Called into Question." *American Dental Association*, August 2, 2016. http://www.ada.org/en/science-research/science-in-the-news/the-medical-benefit-of-daily-flossing-called-into-question.

Merton, Robert. "Science and the Social Order." *Philosophy of Science* 5, no. 3 (1938): 321–37.

Michaels, David. *Doubt Is Their Product: How Industry's Assault on Science Threatens Your Health*. 1st edition. Oxford: Oxford University Press, 2008.

Mnookin, Seth. *The Panic Virus: The True Story behind the Vaccine-Autism Controversy*. 1st edition. New York: Simon and Schuster, 2012.

Moses-Kolko, Eydie L., Julie C. Price, Nilesh Shah, Sarah Berga, Susan M. Sereika, Patrick M. Fisher, Rhaven Coleman, et al. "Age, Sex, and Reproductive Hormone Effects on Brain Serotonin-1A and Serotonin-2A Receptor Binding in a Healthy Population." *Neuropsychopharmacology: Official Publication of the American College of Neuropsychopharmacology* 36, no. 13 (December 2011): 2729–40. https://doi.org/10.1038/npp.2011.163.

Moslehzadeh, Kaban. "Silness-Löe Index," September 29, 2010. /CAPP/Methods-and-Indices/Oral-Hygiene-Indices/Silness-Loe-Index/.

Muller, H. J., F.A.E. Crew, C. D. Darlington, J.B.S. Haldane, C. Harland, L. T. Hogben, J. S. Huxley, et al. "Social Biology and Population Improvement." *Nature* 144 (September 16, 1939): 521–22. https://doi.org/10.1038/144521a0.

Musgrave, Alan. "The Ultimate Argument for Scientific Realism." In *Relativism and Realism in Science*, edited by Robert Nola. Berlin: Springer Science and Business Media, 1988.

Nestle, Marion, Mark Bittman, and Neal Baer. *Soda Politics: Taking on Big Soda*. 1st edition. Oxford: Oxford University Press, 2015.

Nordhaus, William D. *The Climate Casino: Risk, Uncertainty, and Economics for a Warming World*. New Haven, CT: Yale University Press, 2015.

O'Connell, Ronan. "The Great Dental Floss Scam: You May Never Need to Floss Again." *Techly*, August 19, 2016. http://www.techly.com.au/2016/08/19/great-dental-floss-scam-may-never-need-floss/.

Oláh, K. S. "The Use of Fluoxetine (Prozac) in Premenstrual Syndrome: Is the Incidence of Sexual Dysfunction and Anorgasmia Acceptable?" *Journal of Obstetrics and Gynaecology* 22, no. 1 (January 1, 2002): 81–83. https://doi.org/10.1080/0144361012010180.

Oreskes, Naomi. "Objectivity or Heroism? On the Invisibility of Women in Science." *Osiris* 11 (1996): 87–113. https://doi.org/10.2307/301928.

———. *The Rejection of Continental Drift: Theory and Method in American Earth Science.* New York: Oxford University Press, 1999.

———. "The Scientific Consensus on Climate Change." *Science* 306 (2004):1686.

———. " 'Fear'-Mongering Crichton Wrong on Science." *San Francisco Chronicle.* Accessed October 13, 2018. https://www.sfgate.com/opinion/openforum/article/Fear-mongering-Crichton-wrong-on-science-2698545.php.

———. *Science on a Mission: American Oceanography from the Cold War to Climate Change.* Chicago: University of Chicago Press, accepted pending revision.

Oreskes, Naomi, Daniel Carlat, Michael E. Mann, Paul D. Thacker, and Frederick S. vom Saal. "Viewpoint: Why Disclosure Matters." *Environmental Science & Technology* 49, no. 13 (July 7, 2015): 7527–28. https://doi.org/10.1021/acs.est.5b02726.

Oreskes, Naomi, and Erik M. Conway. *Merchants of Doubt: How a Handful of Scientists Obscured the Truth on Issues from Tobacco Smoke to Global Warming.* Reprint edition. New York: Bloomsbury Press, 2011.

Paul, Diane B. "Eugenic Anxieties, Social Realities, and Political Choices." *Social Research* 59, no. 3 (1992): 663–83. https://doi.org/10.2307/40970710.

———. *Controlling Human Heredity, 1865 to the Present.* Humanities Press, 1995.

Paul, Diane B., and Hamish G. Spencer. "Did Eugenics Rest on an Elementary Mistake?" In *Thinking about Evolution: Historical, Philosophical, and Political Perspectives,* edited by Rama S. Singh and Costas B. Krimbas. Cambridge: Cambridge University Press, 2001.

Paul, Diane B., John Stenhouse, and Hamish G. Spencer. *Eugenics at the Edges of Empire: New Zealand, Australia, Canada and South Africa.* Springer, 2017.

Pearce, Warren, Reiner Grundmann, Mike Hulme, Sujatha Raman, Eleanor Hadley Kershaw, and Judith Tsouvalis. "Beyond Counting Climate Consensus." *Environmental Communication* 11, no. 6 (July 23, 2017): 1–8. https://doi.org/10.1080/17524032.2017.1333965.

Perry, Mark. "For Earth Day: Michael Crichton Explains Why There Is 'No Such Thing as Consensus Science.' " *American Enterprise Institute,* April 20, 2015. http://www.aei.org/publication/for-earth-day-michael-crichton-explains-why-there-is-no-such-thing-as-consensus-science/.

Pope Pius XI. "Casti Connubii: Encyclical of Pope Pius XI on Christian Marriage to the Venerable Brethren, Patriarchs, Primates, Archbishops, Bishops, and Other Local Ordinaries Enjoying Peace and Communion with the Apostolic See." Encyclical, December 31, 1930. https://w2.vatican.va/content/pius-xi/en/encyclicals/documents/hf_p-xi_enc_19301231_casti-connubii.html.

Porter, Theodore M. *Trust in Numbers*. Reprint edition. Princeton, NJ: Princeton University Press, 1996.

Proctor, Robert N. *Racial Hygiene: Medicine under the Nazis*. Cambridge, MA: Harvard University Press, 1988.

———. *Golden Holocaust: Origins of the Cigarette Catastrophe and the Case for Abolition*. 1st edition. Berkeley: University of California Press, 2012.

Proctor, Robert N., and Londa Schiebinger, eds. *Agnotology: The Making and Unmaking of Ignorance*. 1st edition. Stanford: Stanford University Press, 2008.

Psillos, Stathis. *Scientific Realism: How Science Tracks Truth*. New York: Routledge, 2005.

Resnick, Brian. "If You Don't Floss Daily, You Don't Need to Feel Guilty." *Vox*, August 2, 2016. https://www.vox.com/2016/8/2/12352226/dental-floss-even-work.

Rettner, Rachael. "Trump Thinks That Exercising Too Much Uses up the Body's 'Finite' Energy." *Washington Post*, May 14, 2017, sec. Health and Science. https://www.washingtonpost.com/national/health-science/trump-thinks-that-exercising-too-much-uses-up-the-bodys-finite-energy/2017/05/12/bb0b9bda-365d-11e7-b4ee-434b6d506b37_story.html.

Roberts, Dorothy. *Killing the Black Body: Race, Reproduction, and the Meaning of Liberty*. New York: Vintage, 1998.

———. *The Ethics of Biosocial Science | The New Biosocial and the Future of Ethical Science*. The Tanner Lectures on Human Values. Princeton, NJ, 2016. https://www.youtube.com/watch?v=NbCyHY9BH7I.

Rosen, Raymond, Roger Lane, and Matthew Menza. "Effects of SSRIs on Sexual Function: A Critical Review: Journal of Clinical Psychopharmacology." *Journal of Clinical Psychopharmacology* 19, no. 1 (February 1999): 67–85. http://journals.lww.com/psychopharmacology/Fulltext/1999/02000/Effects_of_SSRIs_on_Sexual_Function__A_Critical.13.aspx.

Rubin, Neal. "At a Loss over Dental Floss." *Detroit News*, 2016. http://www.detroitnews.com/story/opinion/columnists/neal-rubin/2016/08/22/rubin-loss-dental-floss/89131294/.

Saint Louis, Catherine. "Feeling Guilty about Not Flossing? Maybe There's No Need." *New York Times*, August 2, 2016, sec. Health. https://www.nytimes.com/2016/08/03/health/flossing-teeth-cavities.html.

Sambunjak, Dario, Jason W. Nickerson, Tina Poklepovic, Trevor M. Johnson, Pauline Imai, Peter Tugwell, and Helen V. Worthington. "Flossing for the Management of Periodontal Diseases and Dental Caries in Adults." *Cochrane Database of Systematic Reviews*, no. 12 (December 7, 2011): CD008829. https://doi.org/10.1002/14651858.CD008829.pub2.

Sanger, Margaret, and H. G. Wells. *The Pivot of Civilization*. Berkshire, UK: Dodo Press, 2007.

Schaffir, Jonathan, Brett L. Worly, and Tamar L. Gur. "Combined Hormonal Contraception and Its Effects on Mood: A Critical Review." *European Journal of Contraception & Reproductive Health Care: The Official Journal of the European Society of Contraception* 21, no. 5 (October 2016): 347–55. https://doi.org/10.1080/13625187.2016.1217327.

Schoenfeld, J. D., and J. P. Ioannides,. "Is Everything We Eat Associated with Cancer? A Systematic Cookbook Review." *American Journal of Clinical Nutrition* 97 (2013): 127–34.

Seaman, Barbara, and Claudia Dreyfus. *The Doctor's Case against the Pill: 25th Anniversary*. Alameda, CA: Hunter House, 1995.

Selleck, Robert. "National Media's Spotlight on Flossing Enables Dental Professionals to Shine." *Dental Tribune*, October 13, 2016. http://www.dental-tribune.com/articles/news/usa/31377_national_medias_spotlight_on_flossing_enables_dental_professionals_to_shine.html.

Shmerling, Robert H. "Tossing Flossing?" *Harvard Health* (blog), August 17, 2016. https://www.health.harvard.edu/blog/tossing-flossing-2016081710196.

Showalter, Elaine, and English Showalter. "Victorian Women and Menstruation." *Victorian Studies* 14, no. 1 (1970): 83–89. http://www.jstor.org.ezp-prod1.hul.harvard.edu/stable/3826408.

"Sink Your Teeth into This Debate over Flossing," *Chicago Tribune*. Accessed August 27, 2017. http://www.chicagotribune.com/news/opinion/editorials/ct-dental-floss-fat-heart-associated-press-edit-0805-jm-20160804-story.html.

Skovlund, Charlotte Wessel, Lina Steinrud Mørch, Lars Vedel Kessing, and Øjvind Lidegaard. "Association of Hormonal Contraception with Depression." *JAMA Psychiatry* 73, no. 11 (November 1, 2016): 1154–62. https://doi.org/10.1001/jamapsychiatry.2016.2387.

Smyth, Chris. "Gum Disease Sufferers 70% More Likely to Get Dementia." *Times*, August 22, 2017, sec. News. https://www.thetimes.co.uk/article/gum-disease-sufferers-70-more-likely-to-get-dementia-alzheimers-rd5xxnxwh.

Spiro, Jonathan Peter. *Defending the Master Race: Conservation, Eugenics, and the Legacy of Madison Grant*. 1st edition. Burlington, VT: University of Vermont Press, 2008.

Spencer, Hamish G., and Diane B. Paul. "The Failure of a Scientific Critique: David Heron, Karl Pearson and Mendelian Eugenics." *British Journal for the History of Science* 31, no. 4 (December 1998): 441–52. https://doi.org/10.1017/S0007087498003392.

Staff. "How a Journalist Debunked a Decades Old Health Tip." Accessed August 27, 2017. http://wrvo.org/post/how-journalist-debunked-decades-old-health-tip.

Staley, Richard. "Partisanal Knowledge: On Hayek and Heretics in Climate Science and Discourse." Presented at the Weak Knowledge: Forms, Functions, and Dynamics, Frankfurt, July 4, 2017. http://www.hsozkult.de/event/id/termine-34489.

Stark, Laura. *Behind Closed Doors: IRBs and the Making of Ethical Research*. 1st edition. Chicago: University of Chicago Press, 2012.

Stern, Nicholas. *The Economics of Climate Change: The Stern Review*. Cambridge: Cambridge University Press, 2007.

———. *Why Are We Waiting?: The Logic, Urgency, and Promise of Tackling Climate Change*. Cambridge, MA: MIT Press, 2015. http://www.jstor.org/stable/j .ctt17kk7g6.

Tello, Monique. "Can Hormonal Birth Control Trigger Depression?" *Harvard Health* (blog), October 17, 2016. https://www.health.harvard.edu/blog/can-hormonal -birth-control-trigger-depression-2016101710514.

Thompson, Kristen M. J. "A Brief History of Birth Control in the U.S." *Our Bodies Ourselves* (blog), 2013. http://www.ourbodiesourselves.org/health-info/a-brief -history-of-birth-control/.

Toufexis, D., M. A. Rivarola, H. Lara, and V. Viau. "Stress and the Reproductive Axis." *Journal of Neuroendocrinology* 26, no. 9 (September 2014): 573–86. https://doi.org /10.1111/jne.12179.

Wagner, Wendy E. "How Exxon Mobil 'Bends' Science to Cast Doubt on Climate Change." *New Republic*, November 11, 2015. https://newrepublic.com/article /123433/how-exxon-mobil-bends-science-cast-doubt-climate-change.

Wang, Amy B. " 'Post-Truth' Named 2016 Word of the Year by Oxford Dictionaries." *Washington Post*, November 16, 2016, sec. The Fix. https://www.washingtonpost .com/news/the-fix/wp/2016/11/16/post-truth-named-2016-word-of-the-year-by -oxford-dictionaries/.

Watkins, Adam. "Why the Male 'Pill' Is Still So Hard to Swallow." *Independent*, 2016. http://www.independent.co.uk/life-style/health-and-families/health-news /why-the-male-pill-is-still-so-hard-to-swallow-a7400846.html.

Weinberg, Steven. *Facing Up: Science and Its Cultural Adversaries*. New edition. Cambridge, MA: Harvard University Press, 2003.

White, Byron. *Stump v. Sparkman*, 435 U.S. 349 (March 28, 1978).

Witkowski, Jan. A., *The Road to Discovery: A Short History of Cold Spring Harbor Laboratory*, New York: Cold Spring Harvard Laboratory Press, 2016.

Witkowski, Jan A., and John R. Inglis. "Davenport's Dream: 21st Century Reflections on Heredity and Eugenics." *Journal of the History of Biology* 42, no. 3 (2009): 593–98.

Wynne, Brian. "May the Sheep Safely Graze? A Reflexive View of the Expert-Lay Knowledge Divide." In *Risk, Environment and Modernity: Towards a New Ecology*, edited by Scott Lash, Bronislaw Szerszynski, and Brian Wynne, 44–83. London: Sage, 1996. http://ls-tlss.ucl.ac.uk/course-materials/GEOGG013 _59466.pdf.

Ziliak, Steve, and Deirdre Nansen McCloskey. *The Cult of Statistical Significance: How the Standard Error Costs Us Jobs, Justice, and Lives.* Ann Arbor: University of Michigan Press, 2008.

Coda

Antonio, Robert J., and Robert J. Brulle. "The Unbearable Lightness of Politics: Climate Change Denial and Political Polarization." *Sociological Quarterly* 52, no. 2 (March 1, 2011): 195–202. https://doi.org/10.1111/j.1533–8525.2011.01199.x.

Baum, Bruce, and Robert Nichols. *Isaiah Berlin and the Politics of Freedom: "Two Concepts of Liberty" 50 Years Later.* London: Routledge, 2013.

Berlin, Isaiah. "Two Concepts of Liberty." *Liberty Reader,* 1958. https://doi.org/10.4324 /9781315091822-3.

Brulle, Robert J. "Institutionalizing Delay: Foundation Funding and the Creation of U.S. Climate Change Counter-Movement Organizations." *Climatic Change* 122, no. 4 (February 1, 2014): 681–94. https://doi.org/10.1007/s10584-013-1018-7.

Daniels, George H. "The Pure-Science Ideal and Democratic Culture." *Science* 156, no. 3783 (June 30, 1967): 1699–1705. https://doi.org/10.1126/science.156.3783 .1699.

Daston, Lorraine J., and Peter Galison. *Objectivity.* New York: Zone Books, 2010.

Deen, Thalif. "U.S. Lifestyle Is Not up for Negotiation." *Inter Press Service News Agency,* May 1, 2012. http://www.ipsnews.net/2012/05/us-lifestyle-is-not-up-for -negotiation/.

Dietz, Thomas. "Bringing Values and Deliberation to Science Communication." *Proceedings of the National Academy of Sciences* 110, no. suppl. 3 (August 20, 2013): 14081–87. https://doi.org/10.1073/pnas.1212740110.

Douglas, Heather. *Science, Policy, and the Value-Free Ideal.* Pittsburgh: University of Pittsburgh Press, 2009.

Dunlap, Riley E., and Robert J. Brulle. *Climate Change and Society: Sociological Perspectives.* Oxford: Oxford University Press, 2015.

England, J. Merton. *A Patron for Pure Science. The National Science Foundation's Formative Years, 1945–57. NSF 82–24,* 1982. https://eric.ed.gov/?id=ED230414.

Fischhoff, Baruch. "The Sciences of Science Communication." *Proceedings of the National Academy of Sciences* 110, no. suppl. 3 (August 20, 2013): 14033–39. https:// doi.org/10.1073/pnas.1213273110.

Fleming, James. *Meteorology in America, 1800–1870.* Baltimore: Johns Hopkins University Press, 2000.

———. *Fixing the Sky: The Checkered History of Weather and Climate Control.* New York: Columbia University Press, 2012.

Greenberg, Daniel S., John Maddox, and Steve Shapin. *The Politics of Pure Science*. Revised edition. Chicago: University of Chicago Press, 1999.

"A Greener Bush." *Economist*, February 13, 2003. http://www.economist.com/node /1576767.

Heilbron, J. L. *The Sun in the Church: Cathedrals as Solar Observatories*. Revised edition. Cambridge, MA: Harvard University Press, 2001.

Herrick, Charles N. "Junk Science and Environmental Policy: Obscuring Public Debate with Misleading Discourse." *Philosophy & Public Policy Quarterly* 21, no. 2/3 (2001): 11–16. http://journals.gmu.edu/PPPQ/article/view/359.

Howe, Joshua P., and William Cronon. *Behind the Curve: Science and the Politics of Global Warming*. Reprint edition. Seattle: University of Washington Press, 2016.

Intergovernmental Panel on Climate Change. "Global Warming of 1.5 °C." *IPCC*, 2018. http://www.ipcc.ch/report/sr15/.

Jamieson, Dale. *Reason in a Dark Time: Why the Struggle Against Climate Change Failed—and What It Means for Our Future*. 1st edition. Oxford: Oxford University Press, 2014.

Jewett, Andrew. *Science, Democracy, and the American University: From the Civil War to the Cold War*. Cambridge: Cambridge University Press, 2012.

Kevles, Daniel J. *The Physicists: The History of a Scientific Community in Modern America, Revised Edition*. Revised edition. Cambridge, MA: Harvard University Press, 1995.

Kohler, Robert E. *Partners in Science: Foundations and Natural Scientists, 1900–1945*. Chicago: University of Chicago Press, 1991.

Leiserowitz, Anthony, and Nicholas Smith. "Knowledge of Climate Change Aacross Global Warming's Six Americas." Yale Project on Climate Change Communication. New Haven, CT: Yale University, 2010. http://climatecommunication.yale .edu/publications/knowledge-of-climate-change-across-global-warmings-six -americas/.

Longino, Helen E. *The Fate of Knowledge*. Princeton, NJ: Princeton University Press, 2001.

McCright, Aaron M., and Riley E. Dunlap. "Challenging Global Warming as a Social Problem: An Analysis of the Conservative Movement's Counter-Claims." *Social Problems* 47, no. 4 (2000): 499–522. https://doi.org/10.2307/3097132.

———. "Social Movement Identity and Belief Systems: An Examination of Beliefs about Environmental Problems within the American Public." *Public Opinion Quarterly* 72, no. 4 (2008): 651–76. http://www.jstor.org.ezp-prod1.hul.harvard.edu /stable/25167658.

Merton, Robert K. "Science and the Social Order." *Philosophy of Science* 5, no. 3 (July 1, 1938): 321–37. https://doi.org/10.1086/286513.

————. *Science, Technology & Society in Seventeenth-Century England*. 1st Howard Fertig paperback edition. New York: Howard Fertig, 2002.

Miller, Howard Smith. *Dollars for Research: Science and Its Patrons in Nineteenth-Century America*. 1st edition. Seattle: University of Washington Press, 1970.

Miller, Kenneth R. *Finding Darwin's God: A Scientist's Search for Common Ground Between God and Evolution*. Reprint edition. New York: Harper Perennial, 2007.

————. *Only a Theory: Evolution and the Battle for America's Soul*. Reprint edition. New York: Penguin Books, 2009.

Mooney, Chris. "How to Convince Conservative Christians That Global Warming Is Real." *Mother Jones* (blog). Accessed August 30, 2017. http://www.motherjones.com/environment/2014/05/inquiring-minds-katharine-hayhoe-faith-climate/.

Moore, Kelly. *Disrupting Science: Social Movements, American Scientists, and the Politics of the Military, 1945–1975*. Princeton, NJ: Princeton University Press, 2009.

National Academies of Sciences, Engineering, Division of Behavioral and Social Sciences and Education, and Committee on the Science of Science Communication: A Research Agenda. *Using Science to Improve Science Communication*. Washington, DC: National Academies Press, 2017. https://www.ncbi.nlm.nih.gov/books/NBK425715/.

Numbers, Ronald L., ed. *Galileo Goes to Jail and Other Myths about Science and Religion*. Reprint edition. Cambridge, MA: Harvard University Press, 2010.

Olson, Scott. "Study: Urban Tax Money Subsidizes Rural Counties." *Indianapolis Business Journal*, 2010. https://www.ibj.com/articles/15690-study-urban-tax-money-subsidizes-rural-counties?v=preview.

Oreskes, Naomi, and Erik M. Conway. *Merchants of Doubt: How a Handful of Scientists Obscured the Truth on Issues from Tobacco Smoke to Global Warming*. Reprint edition. New York: Bloomsbury Press, 2011.

————. *The Collapse of Western Civilization: A View from the Future*. New York: Columbia University Press, 2014.

Oreskes, Naomi, and John Krige, eds. *Science and Technology in the Global Cold War*. Cambridge, MA: MIT Press, 2014.

Pope Francis. *Encyclical on Climate Change and Inequality*. Melville Press, 2015. https://www.mhpbooks.com/books/encyclical-on-climate-change-and-inequality/.

Posner, Richard A. *A Failure of Capitalism: The Crisis of '08 and the Descent into Depression*. Unknown edition. Cambridge, MA: Harvard University Press, 2011.

Proctor, Robert. *Value-Free Science?: Purity and Power in Modern Knowledge*. Cambridge, MA: Harvard University Press, 1991.

Prothero, Stephen. *Religious Literacy: What Every American Needs to Know—And Doesn't*. Reprint edition. New York: HarperOne, 2008.

Reeder, Richard, and Faqir Bagi. "Federal Funding in Rural America Goes Far Beyond Agriculture." *USDA ERS*, 2008. https://www.ers.usda.gov/amber-waves/2009/march/federal-funding-in-rural-america-goes-far-beyond-agriculture/.

Shapin, Steven. *A Social History of Truth: Civility and Science in Seventeenth-Century England*. 1st edition. Chicago: University of Chicago Press, 1995.

"Shaping Tomorrow's World: Our Values." Accessed August 30, 2017. http://www.shapingtomorrowsworld.org/values4stw.htm.

Siegrist, Michael, George T. Cvetkovich, and Heinz Gutscher. "Shared Values, Social Trust, and the Perception of Geographic Cancer Clusters." *Risk Analysis* 21, no. 6 (2001): 1047–54. http://onlinelibrary.wiley.com/doi/10.1111/0272-4332.216173/full.

Stern, Nicholas. *The Economics of Climate Change: The Stern Review*. Cambridge: Cambridge University Press, 2007.

Webb, Brian S., and Doug Hayhoe. "Assessing the Influence of an Educational Presentation on Climate Change Beliefs at an Evangelical Christian College." *Journal of Geoscience Education* 65, no. 3 (August 1, 2017): 272–82. https://doi.org/10.5408/16-220.1.

Weber, Max. "Science as a Vocation." *Daedalus* 87, no. 1 (1958): 111–34. https://doi.org/10.2307/20026431.

Zycher, Benjamin. "The Absurdity That Is the Paris Climate Agreement." *American Enterprise Institute*, May 25, 2017. http://www.aei.org/publication/the-absurdity-that-is-the-paris-climate-agreement/.

Chapter 3

Aaserud, Finn. "Sputnik and the 'Princeton Three': The National Security Laboratory That Was Not to Be." *Historical Studies in the Physical and Biological Sciences* 25, no. 2 (1995): 185–239.

Brewer, P. *From Fireplace to Cookstove: Technology and the Domestic Ideal in America*. Syracuse, NY: Syracuse University Press, 2000.

Bridger, S. *Scientists at War: The Ethics of Cold War Weapons Research*. Cambridge, MA: Harvard University Press, 2015.

Busch, J. "Cooking Competition: Technology on the Domestic Market in the 1930s." *Technology and Culture* 24, no. 2 (April 1983): 222–45.

Cartwright, Nancy. *Hunting Causes and Using Them: Approaches in Philosophy and Economics*. Cambridge: Cambridge University Press, 2007.

Cloud, J. "Imaging the World in a Barrel: CORONA and the Clandestine Convergence of the Earth Sciences." *Social Studies of Science* 31, no. 2 (April 2001): 231–51.

Cowan, R. C. *More Work for Mother: The Ironies of Household Technology from the Hearth to the Microwave*. New York: Basic Books, 1983.

Crawford, E. "Internationalism in Science as a Casualty of the First World War." *Social Science Information* 27 (1988): 163–201.

Dobbs, Betty Jo Teeter. *The Foundations of Newton's Alchemy or "The Hunting of the Greene Lyon."* Cambridge: Cambridge University Press, 1975.

Edgerton, David. *The Shock of the Old Technology and Global History since 1900.* New York: Oxford University Press, 2006.

Engber, Daniel. "The Grandfather of Alt-Science: Art Robinson Has Seeded Scientific Skepticism within the GOP for Decades. Now He Wants to Use Urine to Save Lives." Fivethirtyeight.com, October 12, 2017, https://fivethirtyeight.com/features/the-grandfather-of-alt-science/.

Engdahl, E., and Lidskog, R. "Risk, Communication and Trust: Towards an Emotional Understanding of Trust." *Public Understanding of Science* 23, no. 6 (2014): 703–17.

Fauque, D.M.E. "French Chemists and the International Reorganisation of Chemistry after World War I." *Ambix* 58, no. 2 (July 2011): 116–35.

Freire, Olival. "Science and Exile: David Bohm, the Cold War, and a New Interpretation of Quantum Mechanics." *Historical Studies in the Physical and Biological Sciences* 36, no. 1 (September 2005): 1–34.

Galison, P. "Removing Knowledge." *Critical Inquiry* 31, no. 1 (Autumn 2004): 229–43.

Gusterson, Hugh. *Testing Times: A Nuclear Weapons Laboratory at the End of the Cold War.* Stanford: Stanford University Press, 1992.

Haraway, D. "Situated Knowledges: The Science Question in Feminism and the Privilege of Partial Perspective." *Feminist Studies* 14, no. 3 (1988): 575–99.

———. *Modest—Witness@Second—Millennium.FemaleMan—Meets—OncoMouse.* New York: Routledge, 1997.

Hofstadter, R. *Anti-Intellectualism in American Thought.* New York: Knopf, 1963.

Mitchell, M. X. *Test Cases: Reconfiguring American Law, Technoscience and Democracy in the Nuclear Pacific.* PhD dissertation, University of Pennsylvania, 2016.

Moore, Kelly. 2008. *Disrupting Science: Social Movements, American Science and the Politics of the Military, 1945–1975.* Princeton, NJ: Princeton University Press, 2008.

Pisano R., and Busati P. "Historical and Epistemological Reflections on the Culture of Machines around the Renaissance: How Science and Technique Work?" *Acta Baltica Historiae et Philosophiae Scientiarum* 2, no. 2 (Autumn 2014).

Probstein, Ronald F. "Reconversion and Non-Military Research Opportunities." *Astronautics and Aeronautics* (October 1969): 50–56.

Radin, J. *Life on Ice: A History of New Uses for Cold Blood.* Chicago: University of Chicago Press, 2017.

Richmond, M. L. "A Scientist during Wartime: Richard Goldschmidt's Internment in the U.S.A. during the First World War." *Endeavor* 39, no. 1 (2015): 52–62.

Sagan, Carl. *The Demon-Haunted World: Science as a Candle in the Dark.* New York: Random House, 1995.

Selya, Rena. *Salvador Luria's Unfinished Experiment: The Public Life of a Biologist in a Cold War Democracy*. PhD Dissertation, Harvard University, 2002.

Shapin, Steven. "What Else Is New? How Uses Not Innovations Drive Human Technology." *New Yorker*. May 14, 2007.

———. "The Virtue of Scientific Thinking." *Boston Review*. January 20, 2015.

Smocovitis V. B. "Genetics behind Barbed Wire: Masuo Kodani, Emigré Geneticists, and Wartime Genetics Research at Manzanar Relocation Center." *Genetics* 187 (2011): 357–66.

Wang, J. *American Science in an Age of Anxiety: Scientists, Anti-Communism and the Cold War*. Chapel Hill: University of North Carolina Press, 1999.

———. "Physics, Emotion, and the Scientific Self: Merle Tuve's Cold War." *Historical Studies in the Natural Sciences* 42, no. 5 (November 2012): 341–88.

Wynne, Brian. *Risk Management and Hazardous Wastes: Implementation and the Dialectics of Credibility*. Berlin: Springer, 1987.

Chapter 4

Descartes, René. *Meditations on First Philosophy* (1641). Cambridge: Cambridge University Press, 1988.

Galilei, Galileo. *Two New Sciences* (1638). Translated by Stillman Drake. Madison: University of Wisconsin Press.

Goodman, Nelson. *Fact, Fiction, and Forecast*. Fourth edition. Cambridge, MA: Harvard University Press, 1983.

Hoffman, Andrew. "Climate Science as Culture War." *Stanford Social Innovation Review*, 2012. https://ssir.org/articles/entry/climate_science_as_culture_war.

Horowitz, Alana. "Paul Broun: Evolution, Big Bang 'Lies Straight from the Pit of Hell.'" *HuffPost*, 2012. http://www.huffingtonpost.com/2012/10/06/paul-broun-evolution-big-bang_n_1944808.html.

Hume, David. *A Treatise of Human Nature* (1739). Oxford: Clarendon, 1978.

———. *An Enquiry Concerning Human Understanding* (1748). Indianapolis: Hackett, 1977.

Kuhn, Thomas. *The Structure of Scientific Revolutions* (1962). 50th anniversary edition. Chicago: University of Chicago Press, 2012.

Lange, Marc. "Would Direct Realism Resolve the Classical Problem of Induction?" *Noûs* 38 (2004): 197–232.

———. "Hume and the Problem of Induction." In *Handbook of the History of Logic, Colume 10: Inductive Logic*, edited by Dov Gabbay, Stephen Hartmann, and John Woods. Amsterdam: Elsevier/North Holland, 2011, 43–92.

———. *Because without Cause: Non-Causal Explanations in Science and Mathematics*. New York: Oxford University Press, 2016.

Laudan, Larry. "A Confutation of Convergent Realism." *Philosophy of Science* 48 (1981): 604–15.

———. "The Demise of the Demarcation Problem." In *Physics, Philosophy, and Psychoanalysis*, edited by Robert S. Cohen and Larry Laudan. Dordrecht: Reidel, 1983, 111–27.

Meli, Domenico Bertoloni. "The Axiomatic Tradition in Seventeenth-Century Mechanics." In *Discourse on a New Method*, edited by Mary Domski and Michael Dickson. La Salle, IL: Open Court, 2010, 23–41.

Newport, Frank. "In U.S., 46% Hold Creationist View of Human Origins." Gallup.com, http://www.gallup.com/poll/155003/Hold-Creationist-View-Human-Origins.aspx.

Palmerino, Carla Rita. "Infinite Degrees of Speed: Marin Mersenne and the Debate over Galileo's Law of Free Fall." *Early Science and Medicine* 4 (1999): 268–328.

Sextus Empiricus. *Against the Logicians*. Translated by R. G. Bury. Loeb edition. London: W. Heinemann, 1935.

Sellars, Wilfrid. "Empiricism and the Philosophy of Mind." In Sellars, *Science, Perception and Reality*. London: Routledge and Kegan Paul, 1963, 127–96.

———. "Some Reflections on Language Games. In Sellars, *Science, Perception and Reality*. London: Routledge and Kegan Paul, 1963, 321–58.

Chapter 5

Beck, Ulrich. *Risk Society: Towards a New Modernity*. London: Sage, 1992.

Edenhofer, Ottmar, and Kowarsch, Martin. "Cartography of Pathways: A New Model for Environmental Policy Assessments." *Environmental Science & Policy* 51 (2015): 56–64.

Intergovernmental Panel on Climate Change. *Climate Change 2014—Mitigation of Climate Change: Contribution of Working Group III to the Fifth Assessment Report of the Intergovernmental Panel on Climate Change,* edited by O. Edenhofer, R. P. Pichs-Madruga, Y. Sokona, E. Farahani, S. Kadner, and K. Seyboth. Cambridge: Cambridge University Press, 2014.

Koch, N., G. Grosjean, S. Fuss, and O. Edenhofer. "Politics Matters: Regulatory Events as Catalysts for Price Formation under Cap-and-Trade." *Journal of Environmental Economics and Management,* 78 (2016): 121–39. doi: 10.1016/j.jeem.2016.03.004.

Kowarsch, Martin. *A Pragmatist Orientation for the Social Sciences in Climate Policy: How to Make Integrated Economic Assessments Serve Society. Boston Studies in the Philosophy and History of Science* 323. Switzerland: Springer International Publishing, 2016.

Kowarsch, Martin, Jason Jabbour, Christian Flachsland, Marcel T. J. Kok, Robert Watson, Peter M. Haas, et al. "A Road Map for Global Environmental Assessments." *Nature Climate Change* 7, no. 6 (2017): 379–82.

Putnam, Hilary. *The Collapse of the Fact/Value Dichotomy and Other Essays*. Cambridge, MA: Harvard University Press, 2004.

Sarewitz, Daniel. "How Science Makes Environmental Controversies Worse." *Environmental Science & Policy* 7, no. 5 (2004): 385–403.

Chapter 6

Baker, Monya. "Biotech Giant Publishes Failures to Confirm High-Profile Science." *Nature* 530, no. 7589 (2016). https://www.nature.com/news/biotech-giant -publishes-failures-to-confirm-high-profile-science-1.19269.

Barone, Michael. "Why Political Polls Are So Often Wrong." *Wall Street Journal*, November 11, 2015. https://www.wsj.com/articles/why-political-polls-are-so-often -wrong-1447285797.

Bem, D. J. "Feeling the Future: Experimental Evidence for Anomalous Retroactive Influences on Cognition and Affect." *Journal of Personality and Social Psychology* 100, no. 3 (2011): 407–25. doi: 10.1037/a0021524.

Bhattacharjee, Yudhijit. "The Mind of a Con Man." *New York Times*, April 26, 2013 .http://www.nytimes.com/2013/04/28/magazine/diederik-stapels-audacious -academic-fraud.html?pagewanted=all.

Carey, Benedict. "Many Psychology Findings Not as Strong as Claimed, Study Says." *New York Times*, August 27, 2015. https://www.nytimes.com/2015/08/28/science /many-social-science-findings-not-as-strong-as-claimed-study-says.html.

Carey, Benedict, and Pam Belluck. "Doubts about Study of Gay Canvassers Rattle the Field." *New York Times*, May 25, 2015. https://www.nytimes.com/2015/05/26 /science/maligned-study-on-gay-marriage-is-shaking-trust.html.

Festinger, Leon. *A Theory of Cognitive Dissonance*. Evanston, IL: Row, Peterson and Company, 1957.

Freedman, David H. "Lies, Damned Lies, and Medical Science." *Atlantic*, November 2010. https://www.theatlantic.com/magazine/archive/2010/11/lies-damned -lies-and-medical-science/308269/.

Freedman, Leonard P., Iain M. Cockburn, and Timothy S. Simcoe. "The Economics of Reproducibility in Preclinical Research." *PLOS Biology* 13, no. 6 (2015). doi: 10.1371/journal.pbio.1002165.

Ioannidis, John P. A. "Why Most Published Research Findings Are False." *PLOS Medicine* 2, no. 8 (2015): e124. https://doi.org/10.1371/journal.pmed.0020124.

Jennions, M. D., and A. P. Møller. "Relationships Fade with Time: A Meta-analysis of Temporal Trends in Publication in Ecology and Evolution." *Proceedings of the Royal Society of London. Series B. Biological Sciences* 269, no. 1486 (2002): 43–48.

John, Leslie K., George Loewenstein, and Drazen Prelec. "Measuring the Prevalence of Questionable Research Practices with Incentives for Truth Telling." *Psychological Science* 23, no. 5 (2012): 524–32.

Koricheva, Julia, and Jessica Gurevitch. "Uses and Misuses of Meta-analysis in Plant Ecology." *Journal of Ecology* 102, no. 4 (2014): 828–44.

LaCour, Michael J., and Donald P. Green. "When Contact Changes Minds: An Experiment on Transmission of Support for Gay Equality." *Science* 346, no. 6215 (2014): 1366–69. doi: 10.1126/science.1256151.

Lehrer, Jonah. "The Truth Wears Off." *New Yorker*, December 5, 2010. https://www.newyorker.com/magazine/2010/12/13/the-truth-wears-off.

Lehrer, Jonah, and Ed Vul. "Voodoo Correlations: Have the Results of Some Brain Scanning Experiments Been Overstated?" *Scientific American*, January 20, 2009. https://www.scientificamerican.com/article/brain-scan-results-overstated/.

Lord, C. G., L. Ross, and M. R. Lepper. "Biased Assimilation and Attitude Polarization: The Effects of Prior Theories on Subsequently Considered Evidence." *Journal of Personality and Social Psychology* 37, no. 11 (1979): 2098–109. http://dx.doi.org/10.1037/0022-3514.37.11.2098.

Mathews, David. "Papers in Economics 'Not Reproducible.'" *Times Higher Education*, October 21, 2015. https://www.timeshighereducation.com/news/papers-in-economics-not-reproducible.

Miller, Arthur G., John W. McHoskey, Cynthia M. Bane, and Timothy G. Dowd. "The Attitude Polarization Phenomenon: Role of Response Measure, Attitude Extremity, and Behavioral Consequences of Reported Attitude Change." *Journal of Personality and Social Psychology* 64, no. 4 (1993): 561–74. http://dx.doi.org/10.1037/0022-3514.64.4.561.

Open Science Collaboration. "Estimating the Reproducibility of Psychological Science." *Science* 349, no. 6251 (2015). http://science.sciencemag.org/content/349/6251/aac4716.

Prinz, Florian, Thomas Schlange, and Khusru Asadullah. "Believe It or Not: How Much Can We Rely on Published Data on Potential Drug Targets?" *Nature Reviews Drug Discovery* 10, no. 712 (2011). doi: 10.1038/nrd3439-c1.

Reicher, Stephen, and S. Alexander Haslam. "Rethinking the Psychology of Tyranny: The BBC Prison Study." *British Journal of Social Psychology* 45 (2006): 1–40. doi: 10.1348/014466605X48998.

Vazire, S., L. J. Jussim, J. A. Krosnick, S. T. Stevens, and S. Anglin. In preparation. "A Social Psychological Model of Suboptimal Scientific Practices." University of California, Davis.

Vul, Edward, Christine Harris, Piotr Winkielman, and Harold Pashler. "Puzzlingly High Correlations in fMRI Studies of Emotion, Personality, and Social

Cognition." *Perspectives on Psychological Science* 4, no. 3 (2009): 274–90. https://www.edvul.com/pdf/VulHarrisWinkielmanPashler-PPS-2009.pdf.

Walter, Patrick. "Call to Arms on Data Integrity." *Chemistry World*, July 18, 2013. https://www.chemistryworld.com/news/call-to-arms-on-data-integrity/6390.article.

Yong, Ed. "A Failed Replication Draws a Scathing Personal Attack from a Psychology Professor." *Discover*, March 10, 2012. http://blogs.discovermagazine.com/notrocketscience/2012/03/10/failed-replication-bargh-psychology-study-doyen/#.Wo4g_JM-dPo.

Zimbardo, Philip. "The Stanford Prison Experiment: A Simulation Study of the Psychology of Imprisonment." Stanford University, Stanford Digital Repository.

Chapter 7

Alberts, Bruce, Ralph J. Cicerone, Stephen E. Fienberg, Alexander Kamb, Marcia McNutt, Robert M. Nerem, Randy Schekman, et al. "Self-Correction in Science at Work." *Science* 348, no. 6242 (June 26, 2015): 1420–22. https://doi.org/10.1126/science.aab3847.

Bishop, Dorothy V. M. "What Is the Reproducibility Crisis in Science and What Can We Do about It?" University of Oxford, August 30, 2005. https://dx.plos.org/10.1371/journal.pmed.0020124.

Bloor, David. *The Enigma of the Aerofoil: Rival Theories in Aerodynamics, 1909–1930.* Chicago: University of Chicago Press, 2011.

Brainard, Jeffrey, and Jia You. "What a Massive Database of Retracted Papers Reveals about Science Publishing's 'Death Penalty.'" *Science*, October 18, 2018. https://www.sciencemag.org/news/2018/10/what-massive-database-retracted-papers-reveals-about-science-publishing-s-death-penalty.

Brandt, Allan. *The Cigarette Century: The Rise, Fall, and Deadly Persistence of the Product That Defined America.* 1st reprint edition. New York: Basic Books, 2009.

———. "Inventing Conflicts of Interest: A History of Tobacco Industry Tactics." *American Journal of Public Health* 102, no. 1 (January 2012): 63–71. https://doi.org/10.2105/AJPH.2011.300292.

Brown, Nicholas J. L., Alan D. Sokal, and Harris L. Friedman. "The Complex Dynamics of Wishful Thinking: The Critical Positivity Ratio." *American Psychologist* 68, no. 9 (December 2013): 801–13. https://doi.org/10.1037/a0032850.

———. "The Persistence of Wishful Thinking." *American Psychologist* 69, no. 6 (September 2014): 629–32. https://doi.org/10.1037/a0037050.

Cook, John, Peter Ellerton, and David Kinkead. "Deconstructing Climate Misinformation to Identify Reasoning Errors." *Environmental Research Letters* 13, no. 2 (2018): 024018. https://doi.org/10.1088/1748-9326/aaa49f.

Cook, John, Stephan Lewandowsky, and Ullrich K. H. Ecker. "Neutralizing Misinformation through Inoculation: Exposing Misleading Argumentation Techniques Reduces Their Influence." *PLOS ONE* 12, no. 5 (May 5, 2017): e0175799. https://doi.org/10.1371/journal.pone.0175799.

Darrah, Thomas H., Robert B. Jackson, Avner Vengosh, Nathaniel R. Warner, Colin J. Whyte, Talor B. Walsh, Andrew J. Kondash, and Robert J. Poreda. "The Evolution of Devonian Hydrocarbon Gases in Shallow Aquifers of the Northern Appalachian Basin: Insights from Integrating Noble Gas and Hydrocarbon Geochemistry." *Geochimica et Cosmochimica Acta* 170 (December 1, 2015): 321–55. https://doi.org/10.1016/j.gca.2015.09.006.

Dugan, Andrew. "In U.S., Smoking Rate Hits New Low at 16%." Gallup.com, July 24, 2018. https://news.gallup.com/poll/237908/smoking-rate-hits-new-low.aspx.

Fanelli, Daniele. "How Many Scientists Fabricate and Falsify Research? A Systematic Review and Meta-Analysis of Survey Data." *PLOS ONE* 4, no. 5 (May 29, 2009): e5738. https://doi.org/10.1371/journal.pone.0005738.

Fang, Ferric C., R. Grant Steen, and Arturo Casadevall. "Misconduct Accounts for the Majority of Retracted Scientific Publications." *Proceedings of the National Academy of Sciences* 109, no. 42 (October 16, 2012): 17028–33. https://doi.org/10.1073/pnas.1212247109.

Forman, Paul. "Behind Quantum Electronics: National Security as Basis for Physical Research in the United States, 1940–1960." *Historical Studies in the Physical and Biological Sciences* 18, no. 1 (1987): 149–229. https://doi.org/10.2307/27757599.

Franta, Benjamin, and Geoffrey Supran. "The Fossil Fuel Industry's Invisible Colonization of Academia." *Guardian*, March 13, 2017, sec. Environment. https://www.theguardian.com/environment/climate-consensus-97-per-cent/2017/mar/13/the-fossil-fuel-industrys-invisible-colonization-of-academia.

Frederickson, Barbara L., and Marcial F. Losada. " 'Positive Affect and the Complex Dynamics of Human Flourishing': Correction to Fredrickson and Losada (2005)." *American Psychologist* 68, no. 9 (December 2013): 822.

Friedman, Harris L., and Nicholas J. L. Brown. "Implications of Debunking the 'Critical Positivity Ratio' for Humanistic Psychology: Introduction to Special Issue." *Journal of Humanistic Psychology* 58, no. 3 (May 1, 2018): 239–61. https://doi.org/10.1177/0022167818762227.

Godlee, Fiona, Ruth Malone, Adam Timmis, Catherine Otto, Andrew Bush, Ian Pavord, and Trish Groves. "Journal Policy on Research Funded by the Tobacco Industry." *Thorax* 68 (2013): 1091.

Gonzales, Joseph, and Corbin A. Cunningham. "The Promise of Pre-Registration in Psychological Research." *American Psychological Association*, August 2015. http://www.apa.org/science/about/psa/2015/08/pre-registration.aspx.

Hill, Austin Bradford. "The Environment and Disease: Association or Causation?" *Proceedings of the Royal Society of Medicine* 58 (1965): 295–300. https://www.edwardtufte.com/tufte/hill.

Kennedy, Caitlin. "Why Did Earth's Surface Temperature Stop Rising in the Past Decade?" NOAA, September 1, 2018. https://www.climate.gov/news-features/climate-qa/why-did-earth's-surface-temperature-stop-rising-past-decade.

Layton, Edwin. "Mirror-Image Twins: The Communities of Science and Technology in 19th-Century America." *Technology and Culture* 12, no. 4 (1971): 562–80. https://doi.org/10.2307/3102571.

Lewandowsky, Stephan, Kevin Cowtan, James S. Risbey, Michael E. Mann, Byron A. Steinman, Naomi Oreskes, and Stefan Rahmstorf. "The 'Pause' in Global Warming in Historical Context: (II). Comparing Models to Observations." *Environmental Research Letters* 13, no. 12 (2018): 123007. https://doi.org/10.1088/1748–9326/aaf372.

Lewandowsky, Stephan, James S. Risbey, and Naomi Oreskes. "The 'Pause' in Global Warming: Turning a Routine Fluctuation into a Problem for Science." *Bulletin of the American Meteorological Society* 97, no. 5 (September 14, 2015): 723–33. https://doi.org/10.1175/BAMS-D-14–00106.1.

———. "On the Definition and Identifiability of the Alleged 'Hiatus' in Global Warming." *Scientific Reports* 5 (November 24, 2015): 16784. https://doi.org/10.1038/srep16784.

Lewandowsky, Stephan, Naomi Oreskes, James S. Risbey, Ben R. Newell, and Michael Smithson. "Seepage: Climate Change Denial and Its Effect on the Scientific Community." *Global Environmental Change* 33 (July 1, 2015): 1–13. https://doi.org/10.1016/j.gloenvcha.2015.02.013.

Lewandowsky, Stephan, Toby Pilditch, Jens Koed Madsen, Naomi Oreskes, and James S. Risbey. "Seepage and Influence: An Evidence-Resistant Minority Can Affect Scientific Belief Formation and Public Opinion." *Cognition*, Forthcoming.

Linden, Sander van der, Anthony Leiserowitz, Seth Rosenthal, and Edward Maibach. "Inoculating the Public against Misinformation about Climate Change." *Global Challenges* 1, no. 2 (2017): 1600008. https://doi.org/10.1002/gch2.201600008.

Linden, Sander van der, Edward Maibach, John Cook, Anthony Leiserowitz, and Stephan Lewandowsky. "Inoculating against Misinformation." *Science* 358, no. 6367 (December 1, 2017): 1141–42. https://doi.org/10.1126/science.aar4533.

Markowitz, Gerald, and David Rosner. *Deceit and Denial: The Deadly Politics of Industrial Pollution*. 1st paperback printing edition. Berkeley: University of California Press, 2003.

———. *Lead Wars: The Politics of Science and the Fate of America's Children*. 1st edition. Berkeley: University of California Press, 2013.

McHenry, Leemon B. "The Monsanto Papers: Poisoning the Scientific Well." *International Journal of Risk & Safety in Medicine* 29, no. 3–4 (January 1, 2018): 193–205. https://doi.org/10.3233/JRS-180028.

Michaels, David. *Doubt Is Their Product: How Industry's Assault on Science Threatens Your Health*. 1st edition. Oxford: Oxford University Press, 2008.

Michaels, David, and Celeste Monforton. "Manufacturing Uncertainty: Contested Science and the Protection of the Public's Health and Environment." *American Journal of Public Health* 95, no. S1 (July 1, 2005): S39–48. https://doi.org/10.2105/AJPH.2004.043059.

Mooney, Chris. "Ted Cruz Keeps Saying That Satellites Don't Show Global Warming. Here's the Problem." *Washington Post*, January 29, 2016. https://www.washingtonpost.com/news/energy-environment/wp/2016/01/29/ted-cruz-keeps-saying-that-satellites-dont-show-warming-heres-the-problem/.

Myers, John Peterson, Frederick S. vom Saal, Benson T. Akingbemi, Koji Arizono, Scott Belcher, Theo Colborn, Ibrahim Chahoud, et al. "Why Public Health Agencies Cannot Depend on Good Laboratory Practices as a Criterion for Selecting Data: The Case of Bisphenol A." *Environmental Health Perspectives* 117, no. 3 (March 2009): 309–15. https://doi.org/10.1289/ehp.0800173.

Nosek, Brian A., and D. Stephen Lindsay. "Preregistration Becoming the Norm in Psychological Science." *APS Observer* 31, no. 3 (February 28, 2018). https://www.psychologicalscience.org/observer/preregistration-becoming-the-norm-in-psychological-science.

Oppenheimer, Michael, Naomi Oreskes, Dale Jamieson, Keynyn Brysse, Jessica O'Reilly, Matthew Shindell, and Milena Wazeck. *Discerning Experts: The Practices of Scientific Assessment for Environmental Policy*. First edition. Chicago: University of Chicago Press, 2019.

Oransky, Author Ivan. "Retracted Seralini GMO-Rat Study Republished." *Retraction Watch* (blog), June 24, 2014. http://retractionwatch.com/2014/06/24/retracted-seralini-gmo-rat-study-republished/.

Oreskes, Naomi. "Reconciling Representation with Reality: Unitisation as an Example for Science and Public Policy." *The Politics of Scientific Advice: Institutional Design for Quality Assurance*, January 1, 2011, 36–53. https://doi.org/10.1017/CBO9780511777141.003.

———. *Science on a Mission: American Oceanography from the Cold War to Climate Change*. Chicago: Chicago University Press, 2020.

Oreskes, Naomi, Daniel Carlat, Michael E. Mann, Paul D. Thacker, and Frederick S. vom Saal. "Viewpoint: Why Disclosure Matters." *Environmental Science & Technology* 49, no. 13 (July 7, 2015): 7527–28. https://doi.org/10.1021/acs.est.5b02726.

Oreskes, Naomi, and Erik M. Conway. *Merchants of Doubt: How a Handful of Scientists Obscured the Truth on Issues from Tobacco Smoke to Global Warming.* Reprint edition. New York: Bloomsbury Press, 2011.

"Oxford Reproducibility Lectures: Dorothy Bishop." *NeuroAnaTody.* Accessed January 1, 2019. http://neuroanatody.com/2017/11/oxford-reproducibility-lectures -dorothy-bishop/.

"Perceptions of Science in America." *The Public Face of Science.* American Academy of Arts and Sciences, 2018. https://www.amacad.org/publication/perceptions -science-america.

Phillips, Lee. "The Female Mathematician Who Changed the Course of Physics— but Couldn't Get a Job." *Ars Technica,* May 26, 2015. https://arstechnica.com /science/2015/05/the-female-mathematician-who-changed-the-course-of -physics-but-couldnt-get-a-job/.

Pickrell, John. "How Fake Fossils Pervert Paleontology." *Scientific American.* Accessed January 14, 2019. https://www.scientificamerican.com/article/how-fake-fossils -pervert-paleontology-excerpt/.

Pope Francis. *Encyclical on Climate Change and Inequality.* Melville Press, 2015. https://www.mhpbooks.com/books/encyclical-on-climate-change-and -inequality/.

Proctor, Robert N. *Value-Free Science?: Purity and Power in Modern Knowledge.* Cambridge, MA: Harvard University Press, 1991.

———. *Golden Holocaust: Origins of the Cigarette Catastrophe and the Case for Abolition.* 1st edition. Berkeley: University of California Press, 2012.

"Retraction Watch." *Retraction Watch.* Accessed January 14, 2019. https:// retractionwatch.com/.

Richardson, Valerie. "Climate Change Whistleblower Alleges NOAA Manipulated Data to Hide Global Warming 'Pause.'" *Washington Times,* February 5, 2017. https://www.washingtontimes.com/news/2017/feb/5/climate-change -whistleblower-alleges-noaa-manipula/.

Risbey, James S., Stephan Lewandowsky, Clothilde Langlais, Didier P. Monselesan, Terence J. O'Kane, and Naomi Oreskes. "Well-Estimated Global Surface Warming in Climate Projections Selected for ENSO Phase." *Nature Climate Change* 4, no. 9 (September 2014): 835–40. https://doi.org/10.1038/nclimate2310.

Risbey, James S., Stephan Lewandowsky, Kevin Cowtan, Naomi Oreskes, Stefan Rahmstorf, Ari Jokimäki, and Grant Foster. "A Fluctuation in Surface Temperature in Historical Context: Reassessment and Retrospective on the Evidence." *Environmental Research Letters* 13, no. 12 (2018): 123008. https://doi.org/10.1088 /1748-9326/aaf342.

Saal, Frederick S. vom, Benson T. Akingbemi, Scott M. Belcher, David A. Crain, David Crews, Linda C. Guidice, Patricia A. Hunt, et al. "Flawed Experimental Design

Reveals the Need for Guidelines Requiring Appropriate Positive Controls in Endocrine Disruption Research." *Toxicological Sciences* 115, no. 2 (June 2010): 612–13. https://doi.org/10.1093/toxsci/kfq048.

Siegel, Donald I., Nicholas A. Azzolina, Bert J. Smith, A. Elizabeth Perry, and Rikka L. Bothun. "Methane Concentrations in Water Wells Unrelated to Proximity to Existing Oil and Gas Wells in Northeastern Pennsylvania." *Environmental Science & Technology* 49, no. 7 (April 7, 2015): 4106–12. https://doi.org/10.1021/es505775c.

Steen, R. Grant, Arturo Casadevall, and Ferric C. Fang. "Why Has the Number of Scientific Retractions Increased?" *PLOS ONE* 8, no. 7 (July 8, 2013): e68397. https://doi.org/10.1371/journal.pone.0068397.

Taylor, James. "Global Warming Pause Extends Underwhelming Warming." *Heartland Institute*, August 8, 2014. https://www.heartland.org/news-opinion/news/global-warming-pause-extends-underwhelming-warming.

Tollefson, Jeff. "Earth Science Wrestles with Conflict-of-Interest Policies." *Nature News* 522, no. 7557 (June 25, 2015): 403. https://doi.org/10.1038/522403a.

Wazeck, Milena. *Einstein's Opponents: The Public Controversy about the Theory of Relativity in the 1920s.* Translated by Geoffrey S. Koby. 1st edition. Cambridge: Cambridge University Press, 2014.

Wolfe, Robert M., and Lisa K. Sharp. "Anti-Vaccinationists Past and Present." *BMJ (Clinical Research Ed.)* 325, no. 7361 (August 24, 2002): 430–32.

Yong, Ed. "Psychology's Replication Crisis Is Running out of Excuses." *Atlantic*, November 19, 2018. https://www.theatlantic.com/science/archive/2018/11/psychologys-replication-crisis-real/576223/.

Afterword

"Age-Specific Relevance of Usual Blood Pressure to Vascular Mortality: A Meta-Analysis of Individual Data for One Million Adults in 61 Prospective Studies." *Lancet* 360, no. 9349 (December 14, 2002): 1903–13. https://doi.org/10.1016/S0140-6736(02)11911-8.

Bastasch, Michael, and Ryan Maue. "Take a Look at the New 'Consensus' on Global Warming." Cato Institute, June 21, 2017. https://www.cato.org/publications/commentary/take-look-new-consensus-global-warming.

Becker, Katie. "10 American Beauty Ingredients That Are Banned in Other Countries." *Cosmopolitan*, November 8, 2016. https://www.cosmopolitan.com/style-beauty/beauty/g7597249/banned-cosmetic-ingredients/.

Cancer Council Australia. "Position Statement—Sun Exposure and Vitamin D—Risks and Benefits—National Cancer Control Policy." *National Cancer Control Policy.* Accessed January 20, 2019. https://wiki.cancer.org.au/policy/Position

statement-_Risks_and_benefits_of_sun_exposure#_ga=2.151372857.1466774130
.1547753555–1991479126.1547753555.

Cancer Council Australia. "SunSmart." *Cancer Council Australia.* Accessed January 20, 2019. https://www.cancer.org.au/policy-and-advocacy/position-statements /sun-smart/.

Carey, Kevin. "A Peek Inside the Strange World of Fake Academia." *The New York Times,* December 22, 2017.

CFR—Code of Federal Regulations, Title 21, Sec. 352.10. Accessed January 19, 2019. https://www.accessdata.fda.gov/scripts/cdrh/cfdocs/cfcfr/cfrsearch.cfm?fr =352.10.

Consensus Development Panel. "National Institutes of Health Summary of the Consensus Development Conference on Sunlight, Ultraviolet Radiation, and the Skin. Bethesda, Maryland, May 8–10, 1989." *Journal of the American Academy of Dermatology,* no. 24 (1991): 608–12.

C-SPAN. *Sen. James Inhofe (R-OK) Snowball in the Senate.* Accessed January 19, 2019. https://www.youtube.com/watch?v=3E0a_60PMR8.

Downs, C. A., Esti Kramarsky-Winter, Roee Segal, John Fauth, Sean Knutson, Omri Bronstein, Frederic R. Ciner, et al. "Toxicopathological Effects of the Sunscreen UV Filter, Oxybenzone (Benzophenone-3), on Coral Planulae and Cultured Primary Cells and Its Environmental Contamination in Hawaii and the U.S. Virgin Islands." *Archives of Environmental Contamination and Toxicology* 70, no. 2 (February 1, 2016): 265–88. https://doi.org/10.1007/s00244-015-0227-7.

Gabbard, Mike, Donna Mercado Kim, Laura Thielen, Les Ihara, Clarence Nishihara, and Brickwood Galuteria. *Relating to Water Pollution, SB2571 SD2 HD2 CD1 §* *(2021).* https://www.capitol.hawaii.gov/Archives/measure_indiv_Archives.aspx ?billtype=SB&billnumber=2571&year=2018.

Ghosh, Amitav. *The Great Derangement: Climate Change and the Unthinkable.* 1st edition. Chicago: University of Chicago Press, 2017.

Gross, Paul R., and Norman Levitt. *Higher Superstition: The Academic Left and Its Quarrels with Science.* Reprint edition. Baltimore: Johns Hopkins University Press, 1997.

"IARC Monographs Volume 112: Evaluation of Five Organophosphate Insecticides and Herbicides." *IARC Monographs.* International Agency for Research on Cancer, March 20, 2015.

Jacobsen, Rowan. "Is Sunshine the New Margarine?" *Outside Online,* January 10, 2019. https://www.outsideonline.com/2380751/sunscreen-sun-exposure-skin-cancer -science.

Jacobson, Louis. "Did Trump Say Climate Change Was a Chinese Hoax?" *Politifact.* June 3, 2016. https://www.politifact.com/truth-o-meter/statements/2016/jun/03 /hillary-clinton/yes-donald-trump-did-call-climate-change-chinese-h/.

Lindqvist, P. G., E. Epstein, K. Nielsen, M. Landin-Olsson, C. Ingvar, and H. Olsson. "Avoidance of Sun Exposure as a Risk Factor for Major Causes of Death: A Competing Risk Analysis of the Melanoma in Southern Sweden Cohort." *Journal of Internal Medicine* 280, no. 4 (2016): 375–87. https://doi.org/10.1111/joim.12496.

Mooney, Chris. "Ted Cruz Keeps Saying That Satellites Don't Show Global Warming. Here's the Problem." *Washington Post*. Accessed January 19, 2019. https://www.washingtonpost.com/news/energy-environment/wp/2016/01/29/ted-cruz-keeps-saying-that-satellites-dont-show-warming-heres-the-problem/.

Oberhaus, Daniel. "Hundreds of Researchers from Top Universities Were Published in Fake Academic Journals." *Vice*, August 14, 2018.

Oppenheimer, Michael, Naomi Oreskes, Dale Jamieson, Keynyn Brysse, Jessica O'Reilly, Matthew Shindell, and Milena Wazeck. *Discerning Experts: The Practices of Scientific Assessment for Environmental Policy*. First edition. Chicago: University of Chicago Press, 2019.

"Oxybenzone—Substance Information." *ECHA*. Accessed January 19, 2019. https://echa.europa.eu/substance-information/-/substanceinfo/100.004.575.

Perez, Marco I., Vijaya M. Musini, and James M. Wright. "Effect of Early Treatment with Anti-Hypertensive Drugs on Short and Long-Term Mortality in Patients with an Acute Cardiovascular Event." *Cochrane Database of Systematic Reviews*, no. 4 (October 7, 2009): CD006743. https://doi.org/10.1002/14651858.CD006743.pub2.

Pomerantsev, Peter. "Why We're Post-Fact." *Granta Magazine* (blog), July 20, 2016. https://granta.com/why-were-post-fact/.

"Predatory Conference." Wikipedia. Accessed April 19, 2019. en.wikipedia.org/w/index.php?title=Predatory_conference&oldid=893730268.

Rudwick, Martin J. S. *The Great Devonian Controversy: The Shaping of Scientific Knowledge among Gentlemanly Specialists*. Chicago: University of Chicago Press, 1988.

Schneider, Samantha L., and Henry W. Lim. "Review of Environmental Effects of Oxybenzone and Other Sunscreen Active Ingredients." *Journal of the American Academy of Dermatology* 80, no. 1 (January 1, 2019): 266–71. https://doi.org/10.1016/j.jaad.2018.06.033.

Smith, Tara C. "Vaccine Rejection and Hesitancy: A Review and Call to Action." *Open Forum Infectious Diseases* 4, no. 3 (July 18, 2017). https://doi.org/10.1093/ofid/ofx146.

"Sunscreen Fact Sheet." *British Association of Dermatologists*. Accessed January 20, 2019. http://www.bad.org.uk/for-the-public/skin-cancer/sunscreen-fact-sheet#sun-safety-tips.

"Tobacco Companies Ordered to Place Statements about Products' Dangers on Websites and Cigarette Packs." Campaign for Tobacco-Free Kids, January 5, 2018. www.tabaccofreekids.org/press-releases/2018_05_01_correctivestatements.

Trump, Donald J. "The Concept of Global Warming Was Created by and for the Chinese in Order to Make U.S. Manufacturing Non-Competitive." Tweet. *@realDonaldTrump* (blog), November 6, 2012. https://twitter.com/realDonaldTrump/status/265895292191248385.

"United States v. Philip Morris (D.O.J. Lawsuit)." Public Health Law Center. Accessed May 29, 2019. http://publichealthlawcenter.org/topics/tobacco-control/tobacco-control-litigation/united-states-v-philip-morris-doj-lawsuit.

Wang, Jiaying, Liumeng Pan, Shenggan Wu, Liping Lu, Yiwen Xu, Yanye Zhu, Ming Guo, and Shulin Zhuang. "Recent Advances on Endocrine Disrupting Effects of UV Filters." *International Journal of Environmental Research and Public Health* 13, no. 8 (August 2016). https://doi.org/10.3390/ijerph13080782.

"Where Is Glyphosate Banned?" *Baum Hedlund Aristei Goldman* (blog), November 2018. https://www.baumhedlundlaw.com/toxic-tort-law/monsanto-roundup-lawsuit/where-is-glyphosate-banned/.

Zurcher, Anthony. "Does Trump Still Think It's All a Hoax?," *BBC News*, June 2, 2017, sec. US and Canada. https://www.bbc.com/news/world-us-canada-40128034.

CONTRIBUTORS

Naomi Oreskes is professor of the history of science and affiliated professor of earth and planetary sciences at Harvard University. An internationally renowned geologist, science historian, and author, Oreskes is a leading voice on the role of science in society and the issue of anthropogenic climate change. She is the author of both scholarly and popular books and articles, including *The Rejection of Continental Drift, Plate Tectonics: An Insider's History of the Modern Theory of the Earth, Merchants of Doubt: How a Handful of Scientists Obscured the Truth on Issues from Tobacco Smoke to Global Warming*, and *The Collapse of Western Civilization: A View from the Future*.

Jon A. Krosnick is Frederick O. Glover Professor in Humanities and Social Sciences and professor of communication, political science, and (by courtesy) psychology at Stanford University. He is director of Stanford's Political Psychology Research Group and research psychologist at the US Census Bureau. He is a fellow of the American Academy of Arts and Sciences and the American Association for the Advancement of Science.

Susan Lindee is the Janice and Julian Bers Professor of the History and Sociology of Science at the University of Pennsylvania. Most recently, she is the author of *Moments of Truth in Genetic Medicine*.

Marc Lange is the Theda Perdue Distinguished Professor of Philosophy at the University of North Carolina, Chapel Hill. Most recently, he is the author of *Because without Cause: Non-Causal Explanations in Science and Mathematics*.

Ottmar Edenhofer is director of the Potsdam Institute for Climate Impact Research, and is professor of the economics of climate change at Technical University Berlin. He is also director of the Mercator Research Institute on Global Commons and Climate Change and adviser to the World Bank.

Martin Kowarsch is head of the working group on Scientific Assessments, Ethics, and Public Policy at Mercator Research Institute. His work focuses on the interface of science, policy, and society, as well as on values and ethics in integrated environmental assessments.

Stephen Macedo is the Laurance S. Rockefeller Professor of Politics and the former director of the University Center for Human Values at Princeton University. His books include *Liberal Virtues: Citizenship, Virtue, and Community in Liberal Constitutionalism, Diversity and Distrust: Civic Education in a Multicultural Democracy*, the coauthored *Democracy at Risk: How Political Choices Undermine Citizen Participation, and What We Can Do about It*, and *Just Married: Same-Sex Couples, Monogamy, and the Future of Marriage* (Princeton). He is a member of the American Academy of Arts and Sciences.

INDEX

tobacco industry (continued)
290n26; paid studies and, 213,
240–42, 292n44; suppressed
information and, 65–66, 266n136
Traditional Chinese Medicine
(TCM), 62
traditions, 37, 43, 60, 62–63, 78,
102, 136, 165, 209, 220, 235, 260n39,
262n70
transparency, 79, 152, 164, 178, 198,
211–12, 227
Trump, Donald: climate science and,
10, 13, 191–92, 194, 293n3; exercise
and, 269n25; petroleum industry
and, 194; policy assessments and,
191–92, 194; social costs of carbon
(SCC) and, 191–92; vaccinations
and, 15–16, 71
trust: birth control and, 104–17; consensus and, 2 (*see also* consensus);
continental drift and, 80–87; crisis
of, 1, 3, 13, 19, 237; dental floss and,
118–27; empiricism and, 24–30;
epistemologies and, 49–59; eugenics and, 87–104; everyday, 9,
163–66, 177–79; faith and, 36, 56, 66,
68, 168, 178, 183; full disclosure and,
66, 237, 243, 269n14, 272n66; great
men and, 3; informed, 4, 60, 141;
Limited Energy Theory and,
76–80; Pascal's Wager and, 140–46;
peer review and, 3, 7, 9, 32, 53, 58, 66,
75, 130–31, 140, 145–46, 182, 221, 237,
240–41, 247, 252, 266n136, 293n55;
policy assessments and, 191–201;
positive knowledge and, 19–23; and
present and future of science, 206;
reliable knowledge and, 127–39;
revisions and, 32, 36, 38, 199–200,

247; science studies and, 39–49;
social effects and, 12, 18, 55, 57–58,
64, 68, 128, 141, 149, 152, 164, 166,
173, 175, 178, 192–93, 222, 225,
246–47, 268n4; values and, 8
(*see also* values)
truth: circularity and, 181; complete
grasp of, 7; consensus and, 222, 249;
criterion for, 183; critical interrogation and, 58–60, 68, 137, 246, 266n136,
284n2; dimensional homogeneity
and, 188–89; disclosure and, 66, 237,
243, 269n14, 272n66; empiricism
and, 24, 32; epistemologies and, 50,
165, 169, 175, 177, 179; eugenics and,
100; innovation and, 208–9;
instability of scientific, 74–75;
intellectual honesty and, 13;
justification and, 7, 24, 26, 42, 68, 71,
95–98, 151, 158, 183–86, 191, 198, 217,
224, 282n26; Limited Energy Theory
and, 79; observation and, 19, 21, 24,
74; peer review and, 3, 7, 9, 32, 53, 58,
66, 75, 130–31, 140, 145–46, 182, 221,
237, 240–41, 247, 252, 266n136, 29n55;
petroleum industry and, 4; positive
knowledge and, 21, 249; post-truth,
70; present and future of science
and, 202, 206, 208; production of,
137; reliable knowledge and, 131, 137,
139; science studies and, 4–5, 46–49;
shared interest in, 5, 68; structural
realism and, 268n12; temporary, 72,
74–75; transparency and, 79, 152, 164,
178, 198, 211–12, 227; utility and, 19,
151, 218–19; values and, 151; Weinberg
on, 269n19
Tuve, Merle, 169–70
Tyson, Charlie, 263n91

The University Center for Human Values Series

Stephen Macedo, Editor